Toward Environmental Wholeness

SUNY series in Environmental Philosophy and Ethics

J. Baird Callicott and John van Buren, editors

Toward Environmental Wholeness

Method in Environmental Ethics and Science

PATRICK H. BYRNE

SUNY
PRESS

Published by State University of New York Press, Albany

Printed in the United States of America

For information, contact State University of New York Press, Albany, NY
www.sunypress.edu

Library of Congress Cataloging-in-Publication Data

Name: Byrne, Patrick H. (Patrick Hugh), 1947– author.
Title: Toward environmental wholeness : method in experimental ethics and science / Patrick H. Byrne.
Description: Albany : State University of New York Press, [2024]. | Series: SUNY series in environmental philosophy and ethics | Includes bibliographical references and index.
Identifiers: LCCN 2023029854 | ISBN 9781438496986 (hardcover : alk. paper) | ISBN 9781438496993 (ebook) | ISBN 9781438496979 (pbk. : alk. paper)
Subjects: LCSH: Environmental ethics. | Climatic changes—Moral and ethical aspects.
Classification: LCC GE42 .B97 2024 | DDC 179/.1—dc23/eng/20231025
LC record available at https://lccn.loc.gov/2023029854

10 9 8 7 6 5 4 3 2 1

To Joseph Flanagan, SJ,
my teacher, mentor, and dear friend.

Contents

Acknowledgments

A great deal of my scholarly work has grown out of my teaching, and the present book is yet another example. Although I have long been committed to environmental and climate change concerns, I only began teaching courses in environmental ethics several years ago. As I delved more deeply into the literature and engaged with my very talented and active students, I came to realize that several important philosophical questions remain contested. I also realized that I had something of value to contribute to those debates on the basis of my recently completed book *The Ethics of Discernment*. In the last two chapters of that book, I proposed a method for ethics based upon the method that Bernard Lonergan developed for theology—namely, his method of eight integrated "functional specialties." I knew that a concrete illustration would be needed to flesh out what in those chapters I proposed only in a general way. Unexpectedly, I discovered in my teaching of environmental ethics not only a good way to illustrate that method, but a field in need of the kinds of clarifications it could bring.

First and foremost, therefore, I owe my many students a debt of gratitude. Their passions, their thirst to learn about environmental ethics and science, their thoughtful writing, and often their pushback on ideas I presented in class inspired me to write this book, and made it a better book. I also wish to thank my colleagues at Boston College, Profs. Holly VandeWall, David Storey, and Conevery Valencius who generously shared with me their own extensive expertise both in the fields of environmental ethics and history, and how to teach in this field.

I am also grateful to the community of scholars of the thought of Bernard Lonergan for their constant encouragement and feedback. I am very grateful to Profs. Thomas McPartland, Patout Burns, Kenneth

Melchin, and Gerard Whelan, SJ, who encouraged me to turn a lengthy essay into a book. I extend special thanks to Profs. Fred Lawrence, Christopher Berger, Kerry Cronin, Patrick Daly, MD, Fr. Giadio DiBiasi, Mary Elliot, Benjamin Hohman, Chae Young Kim, Timothy Muldoon, Donna Perry, Matthew Petillo, Ligita Ryliškytė, SJE, and Jeremy Wilkins, all of whom contributed many valuable comments on my work in progress. Doctoral candidate Sean Haefner did a thorough proofread of the final draft and his numerous comments improved the text immensely.

Most of all, my deepest thanks to my wife Joan for her love, patience, and support. Without her, this book would not have been possible.

Introduction

Environmental Ethics as Historical

What are the right actions to take regarding our natural environment? In particular, what should be done in response to the ongoing findings of scientists regarding the environment and climate change? What ethical principles should guide us in answering these and related questions? We live in a time of intense conflict regarding the right courses of action to take regarding the natural environment and the changing climate. While there is now a widespread consensus that substantial new actions must be taken, there are also strong and influential dissenting voices. How are we to sort through these intense conflicts over the right course forward?

Over the past several decades, these conflicts played out intensely in several parts of the world, including presidential politics in the United States. In September of 2016 President Barack Obama committed the United States to join the other 195 signatories to the UN Paris Agreement to reduce greenhouse gas emissions.[1] Shortly after his inauguration a year later, President Donald Trump reversed that commitment, announcing that the US would withdraw from the Paris Agreement. In addition to withdrawing from the Paris Agreement, the Trump administration took many other steps to block or reverse initiatives that would have helped to realize this goal, as well as numerous other environmental protections. These policies were in turn again reversed by his successor, President Joseph Biden.

It would be a mistake, however, to assume that the election of Joseph Biden has resolved the conflicts. Climate change denial and opposition to environmental protection did not begin with Donald Trump. His opposition to environmental and climate control policies, and even his publicly expressed doubts about the reality of climate change and about the role of human causation in it, appealed to large numbers of voters and played a major role in his election. Trump appealed to a significant portion of the electorate in the United States that had been profoundly influenced by a widespread network of climate change denial, which had been growing well before the election of 2016. (For further details, see chapters 15, 16, and 17.)

Although the election of President Biden, the US return to the Paris Agreement, and the reversal of many of the Trump-era policies were important events favoring one strong side of the conflict, the underlying culture of climate change denial and opposition to environmental protections has not gone away. These conflicts need to be addressed on many levels. While I personally side with those who affirm the reality of the grave threats to our planet and the ethical urgency for appropriate actions, this book contends that it is just as important to attain critical understanding of the underlying ethical reasons for these claims. This book therefore endeavors to address some but certainly not all of the deeper philosophical and cultural issues that would contribute to attaining this critical understanding.

I will argue that knowledge of the history of ethical and scientific thought is essential to attaining that critical understanding of the foundations needed for ethical responses to environmental and climate change. This claim may sound puzzling if not preposterous to many readers. What could history have to do with environmental and climate change ethics—or any form of ethics, for that matter? After all, it seems that history is about what happened in the past, about things long dead and gone. Ethics, and especially environmental ethics, is about what needs to be done now. Present environmental crises are so serious that we do not have time to waste on such antiquarian curiosities. In fact, if people did know all the complexities of the history of environmental scientific and ethical thought, would it not only confuse them and enervate campaigns that press for political action?

But history is not dead in the past. Countless thoughts, values, beliefs, and actions from the past continue to influence us today. We inherit the achievements of our predecessors as well as their biases and the consequences of their misdeeds. It is important, therefore, to understand the many factors from the past that live on in us today. In this regard I am in fundamental agreement with historian Spencer Weart, who writes that only by "following how scientists in the past fought their way through the uncertainties of climate change, [can we] judge why they speak as they do today."[2] There is a widespread consensus among scientists today about the scientific facts of environmental degradation and climate change. But that consensus did not always exist. Their consensus rests upon the history of more than a century of scientific research. In order to have a solid grounding for ethical actions based on these scientific advances, it is therefore essential to have some grasp of how scientists arrived at them.

Another objection may be raised against the idea that history of ethical thought and of environmental science are essential to environmental ethics. The historical study of ethics, it may be objected, will rob it of its austere moral authority. After all, it seems that moral claims (environmental and otherwise) must be the same for all times and places, and this has certainly been the position of some of the most important ethical thinkers. History seems to reduce ethics to ethical opinions that happened to be held in different historical periods and different cultural settings, thus undermining the normative claims of ethics even in the present. If ethical claims are no more than the products of their own historical epoch, then even present ethical norms must be just as relative as those of the past, and like them, present ethical claims will soon pass away. Why, then, should they be taken seriously, as ethically compelling?

I offered a general answer to such questions in the last two chapters of my book, *The Ethics of Discernment*. There I proposed a method for ethics based upon the method that Bernard Lonergan developed for theology—namely, his method of eight integrated "functional specialties."[3] This is a method for critically evaluating what we have received from the past—especially the deep historical sources of our contemporary conflicts—and for discerning the best courses of action heading into the future. It is a method in which the critical study of history and its conflicts plays a fundamental role. At the time I wrote *The Ethics of Discernment*, I knew that a concrete illustration would be needed to flesh out what in those chapters I proposed only in a general way. This book is intended to provide that concrete illustration by applying that method to the field of environmental and climate change ethics. In doing so, I intend to show why it is of ethical importance to have a critical understanding of the history of environmental and climate change science, as well as of the developing ethical thought about environmental issues.

There is a prejudice that we have inherited from Enlightenment thinking, a prejudice especially prominent in American culture. It is the prejudice that one ought to think for oneself, and not rely on the ideas of others, especially those from the past.[4] Though many Americans will readily regard this prejudice as a normative ideal, we must not forget that it blinds us to an important reality: we always rely on some ideas that we received from others when we deliberate, whether we recognize this or not. We may be intently resisting the influence of some past ideas, but inevitably we will fail to notice the more subtle influences of others. Even the people who claim to be freely thinking for themselves are in fact tacitly relying on unacknowledged ideas from the Enlightenment about what it means to be a "free thinker." No one is free of prejudices and traditions, good or bad; they are just more or less implicit in our thought and action. Critical thinking and authentic ethical actions, therefore, do not come about by completely detaching oneself from all thoughts inherited from the past, for this is impossible. Authentic thinking and action require not separation from but critical evaluation of what one has inherited from the past.

This is the task for which Bernard Lonergan developed his method of eight functional specialties. His method is really an extension and a self-critical refinement of what people already do spontaneously as they evaluate ideas from the past, although they do not always do so carefully or without bias. All of our thinking—whether ordinary commonsense, scientific, or thinking about matters of ethics—utilizes ideas we have absorbed from others who came before us. Among other things, people act toward the future on the basis of what they believe about the past, beliefs that are largely inherited from others. This means that both environmental science and environmental ethics are intrinsically historical, whether we wish this to be the case or not. We inherit both scientific and ethical ideas and attitudes from those who went before us. Often we absorb them unreflectively and rely upon them in the decisions and actions we take. Scientists utilize what they have learned from their predecessors as they conduct their research. Ordinary people use ideas they inherited in their ethical reflections about how to conduct their daily lives. Both scientists and nonscientists alike inherit a mixture of ideas and values, some correct and others erroneous, often without explicit awareness that this is happening. Hence a method that operates with critical self-awareness of this historicity is essential to achieve genuine ethical normativity. This is a book about what we have inherited and the ways it has and has not affected our ethical deliberations and actions with regard to the environment.

In *The Ethics of Discernment* I argued that authentic ethics is the attentive, intelligent, reasonable, responsible, and loving response to what is known about the facts of our concrete circumstances. This means that authentic ethical action cannot ignore objective factual knowledge of real situations. Science contributes mightily to objective knowledge of facts; hence, exactly what is known scientifically about our natural environment and climate change, the degree of probability of that knowledge, and the time it became known are all essential to environmental ethics. This is one more reason why the history of environmental and climate change science is crucial to ethics.

Knowledge of history is also relevant to environmental and climate change ethics for still another reason. Ethics itself evolves. Yet its evolution is not a matter of pure progress. The history of ethics is permeated by conflicts about what is good and right. The history of conflicting ethical views bequeaths an evolving set of conflicting norms from one generation to the next. Therefore, if actions are to be ethical, they have to rest upon critical assessment not only of historically evolving scientific knowledge, but also critical assessment of the evolving and conflicting norms that have been handed down and form the basis for thinking and deciding about ethical responses.

Lonergan's method endeavors to make us more self-aware and make more effective our nonmethodical, spontaneous evaluation and use of inherited ideas. This is to say, he intended to make ordinarily critical evaluation more methodical.[5]

Like ethics in general, the history of environmental and climate change ethics is fraught with conflicts. Conflicts arise in the public domain as well as among scientists themselves. The conflicting versions that people believe about their past histories affect whom they trust and whom they distrust in the present. In addition, conflicts also arise among historians who study and endeavor to understand and document the historical progressions of thought, action, and conflict among laypersons and scientists alike. This means that any adequate method of environmental ethics would need to address the problem of the persistence of conflict in history. Lonergan's method does this, not in the functional specialty of History itself, but in the specialties of Dialectic and Foundations that are connected to it.[6]

The following is a very brief outline of this book. By relying on Lonergan's method and the works of numerous historians, I will show how, over the course of history, scientific and ethical thought about the environment and the changing climate have run up against two moments of extreme crisis that have left us at seemingly irresolvable impasses. The first is a crisis of how to properly understand and value the environment as a whole. The second crisis is the denial of scientific knowledge about climate change.

First, scientific and ethical thought about the environment have pushed beyond the limitations of certain utilitarian frameworks toward a more holistic approach. I will show that the important task of properly formulating this more holistic framework has proven exceedingly difficult and has been fraught with conflicting opinions. I will argue that the difficulties of formulating a coherent articulation of this holistic approach has led to many of the divisions and conflicts in our present situation—dividing those on the "left" who are deeply committed to preserving a pristine nature from human encroachment from those on the "right" who advocate meeting human wants and needs in an economically efficient way. This book intends both to trace the movement that led up to this crisis and to offer a more adequate formulation of holism by drawing on Lonergan's own work in Foundations. This formulation will show the integral connections among the values of both human and nonhuman nature.

Second, I will also show that the history of human thought about climate change ran up against a grave roadblock in a controversy regarding uncertainty and probability in scientific knowledge. This philosophical problem severely undermined efforts to address the perils of climate change in profound and widespread ways. Once again I will offer contributions from Lonergan's own Foundational work regarding probable knowledge that offer a way beyond this impasse.

Rather than delaying direct engagement with the history of environmental science and ethical frameworks, I have deferred a detailed, technical discussion of Lonergan's method to an appendix: "A Method for Environmental Ethics." I trust that the concrete applications of that method in the main chapters of this book will bring clarity to that technical exposition in the appendix, and that the technical exposition will answer questions about the method that arise out of the body of the main chapters.

It is my hope that the approach in this book provides something of value to two different kinds of readers. First, for those concerned with ethical responses to the challenges presented by our changing environments, I offer some new ideas that integrate many of the previous advances in thinking about environmental ethics and environmental science. I also engage some of the crucial conflicts, identify their most fundamental roots, and point to ways toward resolutions. Second, for those interested in Lonergan's method, it is my intention to provide a model of how it can be applied and adapted in ways that are beneficial to fields besides theology.

This book is divided into two parts. Part 1 concerns the history and evaluation of environmental ethics and science. Part 2 focuses on the history and evaluation of the science and ethics of climate change.

By "environmental ethics and science," I mean the human endeavors to understand and act responsibly with regard to limited terrestrial domains, such as bodies of water, the air, areas of land, forests, prairies, tundras, ecosystems of species of plants and animals in limited regions, and the human dependencies upon and interactions with each of these. Such are the topics of part 1.

Ethical thought about these matters has been part of human history from time immemorial. But with the advent of the Industrial Revolution, human impacts upon environments began to accelerate and to raise new kinds of questions for scientific study. The results of these scientific studies, in turn, have begun to raise new kinds of ethical questions about human impact on climates.

While environments are specific and local, climate is global. It is not limited to the environments of particular regions or domains or ecosystems. The climate in one area of the planet is dynamically linked to climate in all other areas. Climate especially concerns the distributions of heat and of water in its liquid, vapor, or frozen forms. As such, climate sets the conditions under which all other environmental domains function. Climate began to change dramatically with the onset of the Industrial Revolution. However, unlike environmental impacts that began to appear almost immediately, the nature of the Industrial Revolution's impacts on climate were initially hidden from view. Scientific knowledge of climate change lagged two centuries behind the reality of its onset. Furthermore, scientific understanding of climate itself has changed dramatically over the past century. As such, it poses a different order of ethical challenge than even that posed by the sciences of environmental impacts. These issues are the focus of part 2.

Environmental ethics, therefore, has to be integrated with climate change ethics. I have chosen to treat climate change ethics second to facilitate the task of integrating these two dimensions—environmental and climate change ethics. The history of environmental ethics reveals the importance of understanding and valuing not merely this or that component of an environment, but environments as wholes that include human as well as nonhuman components. The history of climate change ethics opens out into understanding and valuing a far more comprehensive and dynamic whole that incorporates all the lesser environmental wholes. Human scientific and ethical thought, therefore, have shown a marked movement toward wholeness.

With the help of Lonergan's method, I will show how an adequate holistic ethics is not only possible but already implicitly operative in decisions being made and actions being taken for and against the environment and the climate. The goal of this book is to uncover, account for, and reverse what is mistaken in our past and current ethical responses to the environment and climate, but it is further meant to identify, affirm, and develop the good that is already being thought and done.

Part 1

Environmental Ethics

Chapter 1

From Scientific Facts to Ethical Values

Preliminaries

Environmental ethics has a history to which a great many thinkers have contributed. It is a long history of wrestling with the complex issues of human responsibility toward the environment. It is a history of advances, blockages, and backsliding. Lonergan's method honors the long, dialectical historical processes to which countless human beings have contributed and continue to do so. His method is designed to retrieve, refine, honor, and communicate these many contributions. It is structured to integrate them into a larger whole with policy and action implications. His method relies upon the work of dedicated historians to retrieve the many advances as well as the conflicts and downfalls. Historians narrate the uneven path that thought and action have taken as humanity has wrestled its way forward, endeavoring to meet the complex challenges of its responsibilities toward the environment. His method also adds its own "Dialectical" contributions to the work of historians, discerning how the deepest conflicts were overcome or showing how to overcome those that continue to hold human thought and action in their tenacious grip.[1]

I will be relying in part 1 primarily on the works of two historians: Joseph DesJardins' book *Environmental Ethics* and also Eugene Cittadino's essay "Ecology and American Social Thought."[2] Cittadino and DesJardins offer studies of the ways many thinkers have contributed to the movements of human thought about the natural environment and our ethical responsibilities to it. Both narrate the historical development of scientific and philosophical thinking concerning ecology over the past 150 years. DesJardins focuses on the development of philosophical thought concerning ethical responsibilities to the environment. Cittadino covers much of the same territory but focuses on the interplay between American cultural motifs and scientific methods of environmental study. He traces the evolving history of the science of ecology as well as the complicated history of its relationships to various forms of environmental activism. In doing so, both authors narrate how later stages in the history of thought about the environment emerge from criticisms of the limitations of earlier stages. Following the model of Lonergan's method, I will survey the contributions of these historians to the development of our understanding of the history of environmental ethics, and I will periodically offer contributions from Lonergan's method of Dialectic to identify and offer resolutions to some of the conflicts and impasses they identify.

Although the terms are often used interchangeably, *ecology*, *ecosystem*, and *environment* have distinct meanings. Strictly speaking, ecology is the name of a relatively new science. The term ecology was first coined by Ernst Haeckel in 1866 and only began to be used in scientific literature toward the end of the nineteenth century.

Just what is the object of study of this science of ecology? This question has received an evolving set of answers over the course of that science's history. Initially ecology was a subdivision of botany, as it moved out of the laboratory and began conducting numerous detailed field studies that focused on the interactions among plants in widely different environments. Ecology expanded gradually and unevenly to include the sets of relationships among plants, animals, humans, and other entities and processes.

Ecosystems are now considered to be the objects of ecological scientific investigations. Ecosystems are instances of what theorists now call "complex dynamic systems" of relationships among biotic and abiotic entities and processes. Complex systems are intricate, dynamic feedback networks that are inherently unpredictable; their widely differing pathways are highly sensitive to minute variations in initial conditions. *Ecosystem* therefore refers to a whole system of interactions among all its constituents.

The term environment, on the other hand, tends to refer to an ecosystem from the viewpoint of some entity within that network (as in "the environment *of* this organism"). But it is also often used to refer more generally to an ecosystem without reference to any particular member, or more generally still to something like "ecosystem of ecosystems."

As a philosopher, DesJardins traces the history of ethical thought regarding human responsibilities toward the environment in the most general sense. He emphasizes that environmental ethics must be historically situated, since both the environment and our thinking about it are developing phenomena, with subsequent thought and actions arising in response to what came before: "But how do we even begin making the right decisions? . . . many of our present environmental challenges are the result of decisions made, not by thoughtless or dishonorable people, but in good faith by previous generations . . . many had beneficial consequences. . . . But these decisions have had devastating consequences as well."[3] His historical narrative is further complicated because environmental ethical thought has had to contend with changing scientific understandings of the environment. DesJardins therefore endeavored to connect the changing understandings of environmental ethics to the evolving science of ecology.

It has become a commonplace that science shows us what our ethical responsibilities toward the environment should be. However as DesJardins astutely points out, the relationships between scientific knowledge and ethical imperatives are far from straightforward: "despite increasing scientific study, disputes remain, and they remain because debates about global warming are not simply about the science and facts. . . . Knowledge about the facts alone does not determine what should be done."[4] He recalls for example that many scientific recommendations actually promoted the widespread use of pesticides and herbicides—until Rachel Carson exposed the one-sidedness of those policies in her monumental *Silent Spring*.[5] Based upon the work of Carson and others, DesJardins shows various ways in which scientific practices do have embedded value commitments, and that frequently these function as unexamined assumptions to the detriment of genuinely ethical decision-making based upon scientific findings.[6]

Both Cittadino and DesJardins focus on the shifts that departed from individualistic ways of thinking about environmental matters and moved toward various sorts of holistic approaches. As they also show, the variety of holistic approaches do not seem consistent with each other and in many cases lack even internal coherence. Their historical studies therefore portray environmental thought as having encountered significant difficulties and reached a kind of impasse. Subsequent chapters of this book provide more detailed analyses of their narratives.

It is more accurate to say that Cittadino and DesJardins provide us with histories of the *dialectic* rather than the *development* of environmental thought, because the conflicts and the current impasses mark processes that are not purely *developmental.* Therefore they do not narrate unremitting progress in the field of environmental ethics. Their histories include the real tensions and conflicts that have arisen, some of which remain with us today. These conflicts and impasses therefore invite a response that draws upon Lonergan's method, which is designed to build upon the work of historians in ways that can help the conversations about environmental ethics to move forward. Therefore I will first provide summaries of the historical accounts set forth by Cittadino and DesJardins. I will then shift to an exercise in what Lonergan called the functional specialty of Dialectic, which is designed to analyze the sources of conflict in human history. To anticipate the main points that I will make in greater detail later in this book:

A. To the historical studies by Cittadino and DesJardins I will add contributions from Dialectic to show that implicitly they have provided histories of movements that lead haltingly toward what Lonergan called moral and intellectual conversion. These implicit conversions were expressed explicitly as various forms of "holism."[7]

B. I will then show how their histories trace the debates and conflicts among ethicists and activists about how to best formulate the meaning of holism in environmental ethics, and the problems that resulted from the inadequacies that troubled the various formulations.

C. Because those problems remain unresolved even to the present, DesJardins' historical narrative ends with the call for a pragmatic pluralism as opposed to truth, which reflects his view of the current state of these discussions and debates. I will argue that his call, however, is itself problematic.

D. Finally I will set forth some key ideas from Lonergan's Foundations that offer ways forward beyond some of the dilemmas and impasses that have arisen in the field of environmental ethics. These include especially his ideas about emergent probability, the emerging good, the structure of the human good, the scale of values, the structure of ethical intentionality, and both intellectual and moral conversion.

Hume's Skepticism about Scientific and Ethical Reasoning

Early in his narrative, DesJardins highlights a dilemma raised by the influential criticisms set forth by David Hume. Hume "concluded that although science can tell us all that we can know about nature, it cannot provide the basis for drawing normative conclusions from nature."[8] In particular, Hume criticized the fallacy of arguing from "is" to "ought."

Hume's claim poses a serious problem for environmental ethics. If Hume were correct, advances in scientific knowledge of facts about the environment and climate change should have no impact whatsoever on human ethical responses. Given Hume's lasting influence, it is therefore important to begin with an extended consideration of Hume's claim, as well as a more adequate account of the relationships between scientific knowledge of facts and ethical knowledge of values and actions.

In *A Treatise of Human Nature*, Hume's analysis led him to conclude that passions, not reason, give rise to action. He therefore refuted prior philosophers who had assumed that morality derives from reason. His statement to the contrary has become a classic: "Since morals, therefore, have an influence on the actions and affections, it follows, that they cannot be derived from reason; and that because reason alone, as we have already proved, can never have any such influence. Morals excite passions, and produce or prevent actions. Reason of itself is utterly impotent in this particular. The rules of morality, therefore, are not conclusions of our reason."[9]

Reason, according to Hume, is a part of what he called "understanding." Understanding is the power by means of which human beings form various kinds of connections among ideas. Ideas are famously defined in contrast with impressions at the very beginning of Hume's *Treatise*: "All perceptions of the human mind resolve themselves into two distinct kinds, which I shall call *Impressions* and *Ideas*."[10] Impressions are distinguished from ideas by the intense "force and liveliness" with which they "strike the mind." Impressions are sensations that come by way of seeing, hearing, touching, tasting, smelling, and so forth, but also include passions, emotions, desires, and aversions. By way of contrast, ideas are "faint images of these in thinking and reasoning"[11]—what Lonergan called "impoverished replicas."[12]

The real focus of Hume's analysis was the formation of combinations of ideas ("complex ideas"), especially when the formation of such ideas loses its way. Although the images in memory are tied rather closely to prior impressions, imagination tends to form ideas by separating the elements in original impressions and freely recombining them in all sorts of ways into faint images. Hume's objective was to criticize the ways these more abstract associations are formed, and especially the unfounded sense of reality that people attribute to them.

When Hume spoke of reasoning, he generally meant the chains of discourse whereby people attempt to establish connections between one idea and another. Especially when chains of reasoning become lengthy, they can too easily lose their way, leading to unwarranted claims (his most famous criticism is that the connection of ideas into cause-effect sequences cannot be derived from impressions). Thus, according to Hume, it was important to establish and apply logic, for the "function of logic" he wrote, "is to explain the principles and operations of our reasoning faculty."[13]

Based upon this prior analysis, Hume applied his understanding of logic to criticize the widespread is/ought fallacy: "In every system of morality, which I have hitherto met with, I have always remarked, that the author proceeds for some time in the ordinary way of reasoning, and establishes the being of a God, or makes observations concerning human affairs; when of a sudden I am surprised to find, that instead of the usual copulations of propositions, *is*, and *is not*, I meet with no proposition that is not connected with an *ought*, or an *ought not*."[14] Hume was of course correct that in a strict, formal deductive system, no term can validly appear in a conclusion that did not appear in at least one of the premises. Thus any moral system would be logically invalid if it purports to ground its moral imperatives ("ought" or "ought not" statements) by means of logical arguments from facts about human nature or other factual statements ("is" or "is not"). Hume's point, then, is well taken: *if* reasoning is taken to be identical with strictly formal logical deduction, then there could be no reasoning to moral or ethical claims or actions solely on the basis of propositions about facts. This provides the basis for his further claim "that morality consists not in any relations, *that are the objects of science*; but if examined, will prove with equal certainty, that it consists not in any *matter of fact*, which can be discovered by the understanding. This is the *second* part of our argument; and if it can be made evident, we may conclude, that morality is not an object of reason."[15]

Such a conclusion raises serious problems not only for the prospect of a rational approach to ethical questions but also to the very idea that scientific knowledge might have relevance to ethical reasoning at all. Hume contended instead that morality is derived from passions and efforts to shape passions rather than derived from reason, scientific or otherwise. Unlike ideas and chains of reasoning about facts, according to Hume, passions *do* have the capacity to incite actions: "Since morals, therefore, have an influence on the actions and affections, it follows, that they cannot be derived from reason. . . . Morals excite passions, and produce or prevent actions."[16] Instead of reason, rules of morality are derived from the passions: "Reason is, and ought only to be the slave of the passions, and can never pretend to any other office than to serve and obey them."[17] Reason can at best have a role in morality only by "informing us of the existence of something" that would excite a passion or by discovering a causal connection that would enable us to carry out actions that would satisfy a passion.[18]

At best, reason can shape some passions (e.g., sympathy) in accord with some goal, such as making a profit. Reason, however, cannot specify that goal. For example, suppose one is reasoning about whether to increase an employee's wage, which would satisfy an immediate passion (e.g., sympathy for the employee). Such reasoning itself might be conducted under the influence of another, less immediate passion, the desire to make a profit, which would trump sympathy. In this case reasoning would be a "slave" to the passion for profit. Reasoning enters only to reveal the fact that increasing wages now out of sympathy would decrease one's profit later. But reasoning will not determine which goal ought to be pursued. Only the longer-term passion for profit itself would lead to attenuation of the more immediate passion (sympathy) once the consequences of the latter for the longer term are understood and felt. "Moral reasoning" in this sense can moderate one's passion (sympathy), but only under the influence of the longer-term passion for profit.

An Alternative Account of Scientific and Ethical Reasoning

Hume's radical separation of knowledge of facts from the nature of moral reasoning depends crucially upon where he started: his assumptions about impressions and ideas and their privileged roles in knowledge of facts. As Lonergan pointed out, however, the reasoning Hume actually employed to write his treatise goes well beyond the kind of reasoning left standing at the end of his critique. "The intelligence and reasonableness of Hume's criticizing were obviously quite different from the knowledge he so successfully criticized."[19] Therefore the inconsistency between Hume's performance and what he thought was reasonable provides the basis for Lonergan's Dialectical critique of his account of scientific and ethical reasoning.

Hume was the greatest champion of what Lonergan called the "exceedingly stubborn and misleading myth" that "knowing is taking a look," along with its many implications concerning reality, objectivity, and knowledge.[20] The certainty (or objectivity) of every idea for Hume is adjudicated on the basis of how closely it connects with impressions. For Lonergan, on the other hand, sense impressions and the sensations of passion are only components in the much larger and more intricate processes of human knowing. According to Lonergan, human knowing is a process or "dynamic structure" (as he called it) consisting of numerous mental activities. That process begins with sensations, but these are quickly and spontaneously followed by questions that arise out of those original impressions (Hume, it is worth noting, gives no attention to the significant role of questions in the processes of reasoning). Questions seek neither more impressions nor their pale replicas in ideas; rather, questions seek insights. Insights have their own distinctive and proper contents, which are dramatically different from those of impressions and ideas. Lonergan coined the term *intelligibility* to name this distinctive and proper content of insights.[21] Intelligibilities are strictly speaking unpicturable and unimaginable. They are not impoverished replicas of impressions; rather, they are contents of consciousness of a wholly different kind. Intelligibilities cannot be imagined or pictured, but they can be understood—in insights.

Yet Lonergan's account of the dynamic process of human reasoning does not stop at insights. Merely having an insight does not guarantee its truth. Insights give rise to still further questions: "Is it so? Is it correct?" Such questions set in motion a "self-correcting process" of reasoning "in which insights reveal their shortcomings by putting forth deeds or words or thoughts, and through that revelation prompt the further questions that lead to complementary insights."[22] Reasoning so conceived is motivated, guided, and constituted *as reasoning* by ever further questions. The questions establish the criteria that answers have to satisfy. Only when all questions relevant to a matter at hand have been satisfied can reasoning advance to an accomplished finale in a judgment of fact.[23] Only then can judgments be regarded as reasonable. While sense impressions do provide important input into the self-correcting process of reasoning, they do not provide the total or most fundamental standard for knowing what really is the case. Questions provide that standard.

Hume held that human beings frequently do a poor job in their chains of reasoning unless they strictly adhere to the rules of formal logic. Lonergan agrees that people often reason poorly, but for him, poor reasoning is not a matter of failing to adhere to the rigid rules of formal logic. Human reasoning instead goes astray by failing to properly attend to and satisfy the demands posed by the questions that guide the reasoning process. In *Insight*, Lonergan offered an analysis of some of the basic "biases" that tend to evade or block questions and thereby interfere with the normal self-correcting process of reasoning.[24] Proper attention to one's own reasoning processes is what he meant by "self-appropriation" and what I have called the "ethics of discernment."

Logic, of course, has an important though limited role to play in reasoning, but rigorous logical deduction cannot move beyond the boundaries of its premises. On the other hand, the self-correcting process of reasoning is the source of new and well-grounded premises precisely because it is guided by questions that lead to new insights and to judgments about their soundness (or not). Questions are the source both of the normativity of reasoning as well as its capacity for self-transcendence. Questions lead beyond any given logical system to new insights and new judgments about those insights that provide new premises. In addition, the self-correcting process of reasoning is more flexible and creative than strictly deductive logic, which is why science, common sense, and ethical reasoning are more appropriately identified as self-correcting processes rather than as "logical."[25]

This account of human reasoning therefore undercuts Hume's claim that morals cannot be derived from facts. It is true enough that "ought" statements cannot be *logically* deduced from "is" statements. If, however, reasoning is a matter of question-guided thinking rather than strict logical deduction, then reasoning can and does lead from knowledge of facts to knowledge of values, including knowledge of ethical or moral values.

In my previous book I explained how Lonergan's account of the process of reasoning to factual knowledge expands into a structure or process of ethical intentionality. That expansion begins when knowledge about the facts of some situation give rise to further questions such as: "What can I do?" "What should I do?" "What value would justify doing that?" and "Will I act on this value?" Each question moves ethical reasoning toward additional kinds of insights, imaginations, further questions, judgments of value, decisions, and actions.[26]

I have also examined the roles played by feelings in this process. I am in some agreement with Hume that feelings do play important roles in ethical reasoning, but I disagree with him that reasoning is the slave to feelings. Feelings give us an indispensable consciousness of values, but feelings alone do not automatically provide objectively correct evaluations of the situations out of which they arise or of ethical courses of action to pursue.

Feelings provide a provisional assessment of the values inherent in contemporary situations, but feelings are not infallible. Feelings about situations are only as good as the factual judgments one has about the situation. If one has a naïve or biased set of imaginations or beliefs about what is the case, then one's felt-valuations correspond to these false pictures rather than to realities.

Again, feelings provide an initial valuation of the various courses of action that we come up with in insights answering "What can be done?" But if the self-correcting process does not follow its proper course by considering further questions and insights into alternative courses of actions, then feelings about the value of one exclusive course of action will be distorted, lacking the additional feelings and further questions that would refine and nuance that valuation and lead to a more objective ethical course of action. Thus, unbiased ethical intentionality will ask about the feelings themselves, their meanings, their implications, their distortions and corruptions. While feelings do contribute an indispensable dimension to ethical reasoning, it is questioning ultimately that guides reasoning proper to objective knowledge in ethical as well as factual matters.

Contrary to Hume's, Lonergan's account of reasoning shows that knowledge of facts, including scientific knowledge of facts, is indispensable to objective ethical knowledge. Yet it also shows that scientific knowledge of facts alone does not automatically settle what courses of action in response will be truly ethical. For this, the expansion of human reasoning into the structure of ethical intentionality is necessary. Reasoning guided by questioning places the evaluations offered by feelings into a fuller and proper context—the context of the good as a whole.[27]

This chapter has presented a first example of how Lonergan's Dialectical criticism can serve the advance of environmental ethics. If Hume were correct about the disjunction between scientific knowledge of facts and ethical reasoning, the advances of science would be wholly irrelevant to environmental ethics. The Dialectical criticism presented in this chapter also showed, as DesJardins points out, that the commonplace about science telling us the right ethical path is overly simplified. Not only must our scientific thinking become ever more sophisticated, but so also must our ethical thinking that extends beyond scientific thinking. The remainder of this book is dedicated to showing how this can be accomplished.

Chapter 2

Environmental Ethics in the Utilitarian Mode

Lonergan's method places special emphasis on understanding the history of ethical traditions that we have inherited as a basis for further Dialectical evaluation and refined courses of action. This chapter therefore continues to follow the historical account offered by DesJardins. While DesJardins acknowledges the ways that the various ethical traditions can address environmental issues,[1] his historical narrative really focuses on the dominance of utilitarianism in the earliest stages of environmental ethics, including attempts to extend utilitarianism into animal welfare and animal rights issues. He then examines the difficulties that this approach encountered, followed by its eventual demise and replacement by holistic approaches. This chapter summarizes his account of utilitarianism and its insufficiency as a framework for environmental ethics.

Utilitarianism and Conservationism

DesJardins' historical study focuses on the dominance of utilitarianism in environmental ethics for over half a century.[2] As he points out, "In many ways, utilitarianism is the unofficial ethical theory of public policy in much of North America and Western Europe and, increasingly, for much global policy as well."[3] The deep influence of utilitarianism on environmental public policy and even on early environmental science in the United States began at the end of the nineteenth century with the rise of the conservation movement.

The first prominent figure in DesJardins' historical narrative is Gifford Pinchot, an early pioneer in the conservation movement. He was one of the original members of the US National Forest Commission (1896) and the Boone and Crockett Club (1897), which was founded by Theodore Roosevelt. In 1898 he became head of the Division of Forestry, later renamed United States Forest Service. In that role he promoted the scientific study of forests, going so far as to found and fund graduate schools dedicated to the scientific study of forests from his family's fortune. He also formed the US Forest Service Rangers, many of whom were trained with advanced scientific understanding of forests. Their job was to conserve the national forests from what had previously been a period of unrestrained, rapacious exploitation by the logging industry. Although initially the reception of the US Forest Service Rangers was rather cold, their reputation dramatically improved as a result of their heroic actions in 1910. Along with four thousand US Army troops (including seven companies from the African American Twenty-Fifth Infantry Regiment, known as the Buffalo Soldiers), the Forest Service Rangers fought the largest wildfire in US history to date. The Great Fire of 1910 burned an area of three million acres (the size of Connecticut) in Idaho, Montana, Washington, and British Columbia. It killed seventy-eight people, mostly firefighters. The heroic feats of the US Forest Service Rangers combined with reports of the fire's devastation led to a powerful public commitment to forest conservation in the US.

The great good accomplished by the early conservation movement in general, and Pinchot's US Forest Service in particular, grew out of its utilitarian foundation. This utilitarian ethical framework is revealed in the mission statement of the US Forest Service as quoted by DesJardins: "The object of our forest policy is not to preserve the forests because they are beautiful . . . or because they are the refuges for the wild creatures of the wilderness . . . but . . . the primary object is the making of prosperous homes."[4] In other words, the mission of the US Forest Service was not *preservation* of the environment but *conservation* of the timber resources because of its utilitarian value of maximizing human benefit.[5] The Forest Service would eventually expand the scope of its mission beyond conservation of timber resources to include conservation of recreational uses of the forests. Pinchot's efforts were an ethical advance over previous, undisciplined exploitation of the forests because he held that the resources of the forest "should benefit all citizens, not just the wealthy few who privately owned vast amounts of property."[6] The forests were conserved to serve the *utilitarian* value of providing housing and meeting other such human interests, wants and needs (e.g., recreational enjoyment). In this way Pinchot and the Forest Service sought to maximize benefit for the greatest number of citizens.

It was precisely this ethical framework that led to the US Forest Service's decades-long policy of suppressing and extinguishing forest fires as its primary mission.[7] This policy was adopted in order to further conserve timber for the sake of meeting human needs in years to come, protecting it from the destruction caused by wildfires.[8]

Utilitarianism and Its Conflicts

Although some kind of implicit utilitarian ethics has long motivated human beings, it received its modern influential articulation in the writings of Jeremy Bentham and his follower John Stuart Mill. Bentham wrote: "It is the greatest happiness of the greatest number that is the measure of right and wrong."[9] This has been called the "Greatest Happiness Principle" or "Utility Principle." Beginning with Bentham, the utilitarian precept of promoting the greatest happiness for the greatest number was interpreted as maximizing the pleasures and minimizing the pains summed over all human individuals. Because utilitarianism so conceived rests upon the foundation of pleasure and pain, it has been dubbed "hedonistic utilitarianism." Hedonistic utilitarianism is not an ethics merely intended to justify the provision of pleasure and the avoidance of pain for individuals alone. It is, rather, a social ethics intended to increase the total sum of pleasures minus the totality of pains throughout a society.

Thus conceived, hedonistic utilitarianism might justify an individual forgoing an individual pleasure if calculations could show that doing so would increase the overall happiness in society. On the surface, then, hedonistic utilitarianism seems a simple and obvious way to determine whether a particular course of action would be ethical. All one needs to do is calculate whether that action would increase or decrease total human pleasure, minus pain.

However, hedonistic utilitarianism runs up against several considerable practical difficulties. The first is posed by the monumental task of calculating the sum total of all pleasures and pains in a whole society. Second, hedonistic utilitarianism confronts the still further problem of calculating whether a given action would increase or decrease that sum.

A third problem, originally posed by Mill in his essay *Utilitarianism*, is that pleasures and pains have a qualitative dimension that defies simple quantification. Although Mill was a strong supporter of the utility principle, he acknowledged the serious difficulty posed by incommensurability of pleasures: "Some kinds of pleasure are more desirable and more valuable than others." Hence Mill thought it would be absurd if the calculation "of pleasures should be supposed to depend on quantity alone."[10] This raises the problem of weighing these qualitative differences in one's calculations. Although Mill thought this problem could be overcome, DesJardins points out that his arguments in *Utilitarianism* have not been convincing. As DesJardins puts it, happiness as hedonistic utilitarians conceive "is difficult to quantify." He continues: "Suppose we take good health to be the social goal we seek to maximize. How can we measure and compare the health consequences of various pollution control decisions? In practice it could become easy to substitute for health some quantifiable considerations, such as life expectancy, infant mortality, injury rate, per capita expenditures on health care."[11] DesJardins shows how the difficulties with hedonistic utilitarianism led to revisions by subsequent thinkers into what is called "preference utilitarianism," which "would direct us to satisfy as many individual preferences as possible."[12] It is certainly easier to determine what people express as their preferences, since these preferences show up in how they spend their money or cast their votes or respond in surveys. In general, this is the direction that the utilitarian approach to environmental public policy and ethics has gone in recent history, since it provides a pragmatic solution to the calculation and measurement problems. This also explains how preference utilitarianism has become the unofficial ethical theory of public policy, since the ways that voters express their preferences determine public policies.

However, the move to preference utilitarianism subtly undermines the original spirit of utilitarianism itself. Bentham and Mill both thought that rooting ethics in pleasure and pain provided an objective ground for deciding the ethical worth of a course of action.[13] The problems of calculation, though, led to a replacement of these real, bodily based criteria by mere subjective preferences. Utilitarian ethics, therefore, has become enmeshed in what Lonergan would call a dialectical struggle because of a deep underlying conflict that it has not been able to overcome by means of its own conceptual resources.

DesJardins continues his narration of environmental ethics' reliance upon a utilitarian foundation with the account of a significant conflict between the environmental policy orientation set in motion by Pinchot and an opposing criticism leveled several decades later by Randal O'Toole. O'Toole argued that conservation of public land resources is best accomplished by free market mechanisms rather than by direct governmental agency management: "My economic research has convinced me that Americans can have all the wilderness, timber, wildlife, fish, and other forest resources they want. Apparent shortages of any of these resources are due solely to the Forest Service's failure to sell them at market prices."[14] O'Toole was applying the classical economic efficiency standard to the management of public forest resources. On that basis he argued that bureaucrats in the Forest Service can never accurately measure all the competing human interests and achieve true balance between the sum total of preference satisfactions and dissatisfactions. He also argued that because the Forest Service is largely financed by timber sales from public lands, this provides an incentive to harvest more timber than the markets demand and to sell it at below-market prices.[15]

Although O'Toole argued against the form that utilitarianism took in the US Forest Service as the overseer of forest resource conservation, he nevertheless held firm to the foundation of preference utilitarianism in posing his response. His critique surfaced an inhouse conflict between utilitarians who agree upon utilitarianism as a foundation but disagree on the best means for realizing its ideals. Later on conflicts of a more fundamental sort surfaced.[16] DesJardins shows how these more fundamental conflicts led to a crisis for the utilitarian foundation itself. He points out, for example, that advocacy for preservation of endangered species cannot be supported by a utilitarian foundation. He cites the controversy over the threat to the habitat of the Spotted Owl posed by logging in Pacific Northwest old-growth forests. Obviously, harm to Spotted Owls fall outside the range of hedonistic utilitarianism, since its plight does not enter into the calculation of human pleasures and pains. Nor does preference utilitarianism provide support for preservation of endangered species. "Because the owl has no known *use* and because it does not contribute to society in any obvious way, a utilitarian calculation might suggest that logging should be allowed."[17] One response has been to simply reassert traditional utilitarian norms and argue that the extinction of the Spotted Owl is no ethical problem at all by utilitarian definition. But those who advocated for the preservation of the Spotted Owl habitat were appealing to a deeper, less well-articulated foundation for ethics, a foundation that already exists at least dimly in the consciousness of all human beings. We will return to this issue in later chapters.

Responsibility for the Future: A Utilitarian Dilemma

DesJardins next turns to another issue of dialectical conflict that was already implicit in the approaches of both Pinchot and O'Toole. While they disagreed about the means, both agreed that the ethical objective was to conserve forest resources for human uses, implicitly assuming that the resources are being conserved for *future* uses.

However, claims about ethical responsibilities for conserving the environment for future generations posed a considerable dilemma for the utilitarian ethical framework. Beginning with Bentham, the utilitarian method required calculating present pleasures and pains as the ground for evaluating proposed courses of action. This means utilitarian theory privileges immediate pleasures and pains, discounting the pleasures and pains of the future.[18] After all, how could one accurately measure, calculate, and sum up all the pleasures and the pains of the future, given that they are unknown in the present? Nor does a shift to preference utilitarianism solve this problem; future preferences are no better known than are future pleasures or pains. Furthermore, how far into the future should such calculations be conducted? For the next year? For the next ten years? For the next one hundred years? As DesJardins puts it, the utilitarian privileging of present pleasures would lead to an ironic outcome: "One implication . . . might require us to maximize the present value [overall "happiness"] of our resources by using them now and discounting their value to later generations. . . . Eventually we would be committed to saying that future people do not count at all."[19] DesJardins discusses several other dilemmas regarding the problem of responsibilities for future generations within a utilitarian framework, as well as the attempts that were made to meet them. For instance, he points out that maximizing happiness would seem to imply a commitment to promoting an increase in population (along with increased resources) so as to increase the sum total of pleasures.

DesJardins does not think this host of problems can be adequately resolved within the limits of the utilitarian ethical framework. "*Environmental philosophers have reached a strong consensus that the narrow worldview of classical economics and preference utilitarianism that underlies it must be rejected.*"[20] The strong consensus that he cites constitutes a crisis for utilitarianism as a foundation for environmental ethics. It is a crisis that has come out of the dialectical history of environmental thought, debate, and action. It is a crisis that calls for a fundamental alternative, a "conversion" in the terms of Pope Francis, Lonergan, and others. In the subsequent chapters of this book, I will be following DesJardins and other historians as they trace the ways that the dialectic of environmental history has grappled with what this fundamental alternative must be.

From these impasses one might be led to conclude that there really is no ethical responsibility to future generations, and many philosophers and policy framers have taken exactly this stance. Such conflicting conclusions, however, only follow from utilitarian assumptions. In fact, most people spontaneously think that they really do have at least some ethical responsibilities to future generations. This often shows up in remarks about what kind of world people want to pass along to their grandchildren. In other words, the spontaneous ethical thinking and acting of many people operate outside of the limitations of utilitarianism. In a later chapter, I will propose that an ethics of discernment does greater justice to spontaneous ways that human beings actually do their ethical thinking and acting, especially when it comes to this issue of ethical responsibility about what kind of environment should be passed along to future generations.[21]

Transitional Stage: Animal Welfare and "Ethical Extensionism"

The next stage in DesJardins' historical narrative focuses on the rise of animal welfare ethics (also referred to as "animal rights" or "animal liberation" ethics), which he portrays as a positive development. Initially ethical concerns about animal welfare arose from revulsions at the mistreatment of domestic animals. But it soon expanded to concerns about how human actions impact all animals, domestic or wild. This stage in environmental ethical thinking produced what he calls "a more radical shift" in ethical thought about the environment away from exclusive focus on human interests and preferences, taking into account animals that also inhabit environments. In DesJardins' telling, however, even animal welfare ethics did adequately meet all the ethical challenges that the environment poses for human responsibility because it still relied upon certain utilitarian assumptions.

Animal welfare ethics is relevant to environmental ethics because a vast portion of the natural environment is constituted by the complex interconnections among animal lives, many of which escape ordinary human notice. There are countless ways that animals constitute the environments that humans inhabit, so when we think about the environment and our ethical responsibilities toward it, we inevitably have to face the issue of ethical responsibilities to animals.

One would expect, says DesJardins, that the natural law/virtue/teleological tradition in ethics, rather than a utilitarian framework, would have much more to offer an ethics that grounds responsibility toward animals. This is because this tradition situates ethics within the wider context of a philosophy of nature. He notes, however, that the two central figures in the natural law tradition, Aristotle and Thomas Aquinas, are disappointing in this regard. Both explicitly state as a doctrine of natural law that animals exist for the sake and use of human beings.[22]

The natural law tradition therefore fails to extend "moral standing" beyond the human realm in ways that take into account moral intuitions about treatment of animals. Along with many other environmental ethicists, DesJardins regards moral standing rather than "nature" as the crucial standard for ethical evaluation, because rights are held to derive from moral standing. In other words, just as humans are said to have rights to life and protection from harm on the basis of the moral standing of human dignity, so too animals should have such rights insofar as they can be regarded as having moral standing. However, argues DesJardins, until very recently "only human beings have moral standing." He therefore sees the failure of the natural law tradition as a failure to sufficiently broaden the category of moral standing to include animals.[23]

DesJardins identifies Joel Feinberg, Peter Singer, and Tom Regan (called "ethical extensionists") as the most prominent contributors to extending this concept of moral standing. All three still maintain the idea of *individual* interests (or preferences) as the basis for moral standing, but they extend the concepts of interests and moral standing to include animals as well as humans.

Singer is perhaps the most famous of all the philosophers of the animal welfare movement. Given that utilitarianism traditionally gave such high priority to human interests, it is perhaps surprising that Singer aligned himself with the utilitarian tradition. But he took his point of departure from a position originally articulated by Bentham: the most salient question to ask in determining whether something has interests and moral standing is "Can they *suffer?*"[24] Singer wrote, "The capacity for suffering and enjoyment is a *prerequisite for having interests at all* . . . at an absolute minimum, an interest in not suffering."[25] This means that all sentient beings can have moral standing. Singer acknowledged that different animals suffer in different ways, and that this suffering provides a basis for comparing interests. For example, one cannot prefer an action that overcomes mere human inconvenience if it causes severe animal suffering. Nevertheless, DesJardins points out that the longstanding problem for utilitarianism of measuring and summing up diverse forms of suffering and pleasure remains a significant sticking point for Singer. In addition, since the capacity to experience suffering cannot be extended beyond either sentience or individuals, Singer's ethical foundation provides no basis for addressing a great many ethical issues posed by other dimensions of natural environments—water, air, species as such (versus individual animals), and plant ecosystems cannot be said to have interests or moral standing according to Singer's criteria.

At about the same time that Singer began to articulate his animal welfare ethics, Feinberg wrote an early, influential essay advancing animal welfare ethics. He attempted to do so without explicitly invoking the utilitarian suffering/pleasure calculus. He proposed that individuals can legitimately be said to have interests if they possess "conative life," that is, "conscious wishes, desires, hopes, direction, growth, and natural fulfillments."[26] He argued that we therefore have duties to refrain from impeding the interests of individual animals (including the taking of their lives) in just the same way that we have duties to respect the rights of human beings to pursue their interests (so long as they do not harm the interests or lives of other human beings). Feinberg further argued that this criterion of conative life can also provide the basis for ethical responsibility to future individuals. However, in Feinberg's view, only higher ("conative") animals have such interests and therefore have moral standing and rights. Lower animals have no rights and can be killed as pests. In addition, neither plants nor species as such nor abiotic entities such as water or land can have moral standing or rights. Thus

Feinberg's attempt to extend moral standing beyond human interests still seems rather arbitrary in relationship to wider environmental concerns.

A few years later Tom Regan endeavored to overcome the limitation of both Singer's utilitarianism and Feinberg's foundation in conative life. He adopted something like a deontological/rights approach to ethics. Regan claimed that the ethical obligations owed to animals do not derive merely from their capacities for suffering. Implicitly drawing upon Kant, Regan argued that the moral standing of animals derives from their "inherent value," which makes them ends in themselves. Unlike Kant, however, he held that the inherent value of animals is not rooted in the capacity of their practical reason and respect for the universality of law. Rather, the inherent value of animals is grounded in the fact that each individual animal has a life, or as he puts it, is the "subject-of-a-life": "To be the subject-of-a-life . . . involves more than being alive and . . . conscious. [It involves having] beliefs and desires, perception, memory, and a sense of the future . . . emotional life together with feelings of pleasure and pain, preference and welfare interests, the ability to initiate action in pursuit of their desires and goals."[27] Therefore anything that is the subject-of-a-life has interests and moral standing. On this basis, Regan condemned the use of animals for food, recreation, sport, hunting and fishing, pets, zoos, and scientific or commercial research.[28] In all of these practices, "we fail to treat individuals who have inherent value with the respect they deserve when we treat them as though they were valuable only as a means to some other end."[29]

In DesJardins' telling, these three animal welfare advocates and many others who have followed their lead made a significant advance in environmental ethics by extending the concept of interests to nonhuman animals. Yet he also narrates the history of several subsequent lines of criticism of these thinkers. For one thing, all three offer foundations that only extend to some higher individual animals but not to the invertebrates, "the little things that run the world," as E. O. Wilson put it.[30] Environments are constituted by all ranges of animals, not just those possessing higher sentience. Environments are also constituted by plants, fungi, bacteria, and other organisms, none of which could be said to have moral standing on the bases proposed by Singer, Feinberg, or Regan.

Moreover, in the work of animal welfare philosophers, only *individual* higher animals can be said to have such interests, because only individuals can experience suffering or conation or have desires or goals. This excludes from ethical consideration nonindividual entities such as species, habitats, land, bodies of water, and the atmosphere. Finally, in spite of the effort to expand ethical foundations beyond the restriction to human beings, all three nevertheless remain subtly anthropocentric. All three begin with criteria—suffering, conation, and subject-of-a-life—that we use to ascribe interests to human beings and then extend them to animals by analogy. These approaches do not identify what is *inherently* valuable about animals as such. Rather, they take human beings as the paradigms of moral standing and then identify what in animals is *like* human beings as the basis for attributing moral standing to them.

DesJardins concludes this stage of his narrative by saying that "*critics have come to believe that the animal welfare movement is not an adequate environmental philosophy*."[31] These conflicts between animal welfare advocates and their critics are part of the history of environmental ethics. This history is what Lonergan called a dialectical process: a progression whose resolution requires something like a conversion from satisfaction (anthropocentric individual interests) *toward the whole of value* (an ecocentric perspective).[32] This is exactly the challenge that confronted the history of environmental ethics, according to DesJardins: "Environmental ethics requires more than a simple concern for individual animals of a certain type. . . . A shift to such holistic and truly nonanthropocentric ethics, however, would require a break from tradition. . . . All of what follows in [his book] implicitly takes up the challenge to 'question the basic assumptions of the age.' "[33]

The subsequent chapters of this book follow the narratives of DesJardins and Cittadino as they trace the attempts to articulate a more holistic foundation for environmental ethics.

Chapter 3

Conversion to the Value of Life as a Whole

In the subsequent chapters of his narrative, DesJardins follows the paths blazed by people who made radical shifts away from a utilitarian-based ethics that privileges either anthropocentrism or sentient animals. His narrative portrays the development of environmental ethics as the abandonment of preference for individuals (whether humans or animals) as the criteria for evaluating actions as ethical in favor of some form of holism as necessary to provide a more adequate foundation for environmental ethics.

Conversion to Biocentric Holism: Albert Schweitzer

DesJardins highlights two individuals who underwent radical shifts to a holistic ethical framework in the area of environmental concerns: Albert Schweitzer (1875–1965) and John Muir (1838–1914). Both Schweitzer and Muir described these shifts as moments of conversion for them.[1] Although Muir was a contemporary and adversary of Pinchot, and although both Schweitzer and Muir wrote about their new ethical perspectives well before Singer, Feinberg, or Regan began their work, DesJardins places his discussions of Schweitzer and Muir later in his historical narrative. He does so because the influence of these holistic thinkers only gained wide acceptance in the latter half of the twentieth century, continuing into the present. Their thought had to overcome the hegemony that utilitarianism (whether limited to humans or extended to animals) had established during the first half of that century. This is probably because the holistic foundation they call for is more difficult to formulate than are the basic principles of utilitarianism. Only once the scientific facts about environmental degradation became more widely known could their calls for a more holistic approach be taken seriously.

Although Schweitzer spent his early years as a Lutheran theologian, he was dissatisfied with both traditional and liberal theological thought. His search for ethical foundations eventually led him to become a medical doctor and to devote much of his life rendering medical service to people in Gabon. Yet soon after his arrival in Africa, he experienced a compelling moment of conversion to a new ethical foundation.

Schweitzer describes in almost mystical terms the moment that this idea came to him. While riding on a barge traveling upriver in Africa, "at the very moment when, at sunset, we were making our way through a herd of hippopotamuses, there flashed upon my mind, unforeseen and unsought, the phrase *reverence for life*."[2]

Schweitzer's novel ethical foundation came about through a conversion to the inherent value of *life as a whole*. It was therefore inclusive of all life—human, animal, and plant—and not restricted either to animals of higher sentience or even to individual organisms. Schweitzer's conversion led him to the conviction that the inherent goodness of life even transcends "the destructive and arbitrary power of nature."[3] In other words, he was well aware of the countless moments of cruelty by animals to one another through his experiences in Africa. Nevertheless, Schweitzer saw that life includes all these moments of cruelty and killing as essential to life's continuation. The whole of life possesses a goodness that transcends and cannot be overcome by such moments of violence.

Schweitzer was convinced that his principle of reverence for life offered the basis for an ethics that would restore the proper relationship between humans and nature. He formulated his principle in the following terms: "I am life which wills to live, in the midst of life which wills to live. . . . [Everyone] who has become a thinking being feels a compulsion to give to every will-to-live the same reverence for life that [they] give to [their] own . . . to preserve life, to promote life, to raise to its highest value of life which is capable of development."[4] Schweitzer was so committed to this egalitarian principle that he even refused to kill mosquitoes. Nevertheless he did admit that sometimes it would be necessary to take life in order to reverence life, as when one animal or plant must be sacrificed to feed some other animals, or if an individual life should be terminated in order to end its suffering.[5] While animal welfare ethics also acknowledges the value of life, it is *individuals* that are valued because they possess life; it is not life as such, life as a whole, that is valued. Schweitzer, on the other hand, recognized life as a whole to have a value that cannot be limited to the individuals that happen to be alive.

Though Schweitzer's principle of reverence for life was a significant step forward, there remain unresolved tensions in his attempts to articulate and live out his conversion to the value of life as a whole. For example, there is an inconsistency in Schweitzer's opposition to killing mosquitoes and his recognition that life as a whole involves the killing of some animals by others. Again, though plants are participants in life-as-a-whole, Schweitzer consumed plants while refraining from eating animals.

DesJardins traces how subsequent philosophers endeavored to find ways of resolving the conflicts in Schweitzer's formulation of his holist principle for ethics. He draws particular attention to the work of Paul Taylor, who moved toward what Lonergan would have called a more encompassing or "explanatory" direction.[6] Taylor rephrases Schweitzer's principle as "respect for nature" in the following terms:

> Once we come to understand [an organism's] life cycle and know the environmental conditions it needs to survive in a healthy state, we have no difficulty in speaking about what is beneficial to it and what might be harmful to it. . . . Even when we consider such simple animal organisms as one-celled protozoa, it makes perfectly good sense to a biologically informed person to speak of what benefits or harms them, what environmental changes are to their advantage or disadvantage.[7]

Taylor therefore moved away from any privileging of higher sentience in individual animals toward a holistic perspective that values "things in their relations to one another."[8] Taylor formulated several ethical principles, such as the principle that humans are members of the community of life and as such have no privileged status. Humans are therefore bound by principles of nonmaleficence and noninterference. In other words, humans may not interfere with the life cycles of other organisms or biotic communities.[9]

However, DesJardins draws attention to problematic features of Taylor's more sophisticated holism. Like Schweitzer, Taylor affirms predatory and feeding behaviors of nonhuman animals, while arbitrarily disapproving the very same behaviors on the part of human beings. As DesJardins puts it, "To say that we ought not to 'interfere with' nature implies that humans are somehow outside of or distinct from nature," including the feeding behaviors of natural organisms.[10] This view is in conflict with Taylor's other principle that human beings are part of the community of life. These conflicts within the biocentric tradition of holism are of the kind that Lonergan's methods of Dialectic and Foundations were designed to address. We will return to this issue later.

Conversions to Ecocentric Holism: John Muir

John Muir's holism also came out of a conversion experience. He was already intrigued by nature as a child, and his explorations of nature intensified during his extensive hikes in the North American wilderness as a young man. Although he was raised in a strict Scottish Presbyterian household, he gradually he came to regard the Book of Nature as a "primary source for understanding God."[11] While his conversion may have been gradual, the peak moment in that process occurred in 1869 when he first entered the Yosemite Valley. In his journal he wrote:

> We are now in the mountains and they are in us, kindling enthusiasm, making every nerve quiver, filling every pore and cell of us. Our flesh-and-bone tabernacle seems transparent as glass to the beauty about us, as if truly an inseparable part of it, thrilling with the air and trees, streams and rocks, in the waves of the sun,—a part of *all nature*, neither old nor young, sick nor well, but immortal. . . . How glorious a conversion, so complete and wholesome it is, scarce memory enough of old bondage days left as a standpoint to view it from! In this newness of life we seem to have been so always.[12]

Through this conversion Muir came to recognize that nature as a whole has inherent value that is irreducible to its values for human uses or even to the values of its individual members. He "defended the spiritual and aesthetic value of wilderness, as well as the inherent worth of other living things."[13] He came to regard the wilderness as a profound source of religious inspiration and a corrective to the corrupting influences of modern life. In this way Muir differs from and expands upon Schweitzer's *biocentric* holism, moving toward a more encompassing *ecocentric* holism.[14]

Muir founded the Sierra Club and, through his friendship with Theodore Roosevelt, was instrumental in establishing Yosemite and other locations as national parks. On the basis of his vision of the intrinsic value of nature as a whole, he argued for preservationism against Pinchot's conservationism. Muir argued against Pinchot that the wilderness should be preserved because of its inherent value, not merely because of its utility to human beings. The intensity of the conflict between their ethical foundations became focused in the debate over building a dam that would flood the Hetch Hetchy Valley, destroying thousands of acres of forests and changing the awe-inspiring view of the valley forever. Conservationist Pinchot supported building the dam because he viewed it as an ethically justifiable use of the environment to meet human needs. Muir opposed it in hopes of preserving the intrinsic value of the valley. He wrote, "Dam Hetch-Hetchy! As well dam for water-tanks the people's cathedrals and churches, for no holier temple has ever been consecrated by the heart of man."[15] Muir lost the battle. President Woodrow Wilson signed into law the legislation authorizing construction of the dam in December 1913. Muir died just over a year later. The dam is a rallying cry for many of his followers and cements his inspiring example for them in their preservationist work.[16]

The "Wilderness Myth" and Its Discontents

Muir's influence grew substantially long after his death, especially among environmental activists in the latter half of the twentieth century. However, DesJardins is critical of at least one strand of environmental thought that has grown out of Muir's vision: its foundational reliance on the "myth of the wilderness."

DesJardins narrates the transformations of the symbol of "wilderness" in American culture. He notes that "wilderness" is a fairly recent and especially Eurocentric concept. Significantly, "wilderness" is usually defined in negative terms; it is generally understood as that which is free from human interference. Wilderness is the place "where the earth and its community of life are untrammeled by man."[17] But the preservationist idea of *setting aside wilderness areas* subtly implies a significant degree of human management, if not interference. The strict separation between humans and nature suggested by "untrammeled" is therefore misleading, though this is unnoticed by many preservationists. DesJardins argues that this has to do with specifically European traditions that have been incorporated into and transformed by American culture.

According to DesJardins, Puritan English settlers in the North American continent initially relied upon their understanding of biblical narratives to interpret the wilderness negatively. In their readings of the story of the expulsion from the Garden of Eden and of biblical desert imagery, the wilderness was regarded as a cruel, harsh, perilous threat, a place to be avoided and feared. The diaries and literature of the earliest English settlers (especially captivity narratives) depict the wilderness as a place where the faith and moral fealty of Puritans are tested, just as Jesus was tempted by the devil in the desert.[18]

Yet in one of the most remarkable shifts in American cultural history, the wilderness was transformed into a positive symbol. Adapting to the environment of the North American continent was not easy for the English settlers. The skills, customs, and practices that the Puritans brought with them were not up to the task of living in the new environments. They gradually began to look to the wilderness and its Native American inhabitants for instruction first on how to survive in a strange new land and then on how to convert the wilderness into a new Garden of Eden. "Proof that the Puritans were indeed the new 'chosen people' would be manifest in how they fared in this wilderness"—that is, how well they dominated and tamed it.[19] This transformed symbolism of the wilderness laid a cultural foundation for a Lockean view of the wilderness as lying out there, ready to be converted by hard work into private property. It also paved the way for the utilitarian ethics that would guide public policy.

Subsequently the symbol of wilderness underwent yet another dramatic shift in American culture, this time under the influence of Jean-Jacques Rousseau and the Romantic movement that succeeded him. In many ways, Romanticism was a reaction against the aggressive taming of wilderness into a garden. Romanticism instead viewed the wilderness itself as the original Garden of Eden, as yet untouched by human work. The wilderness thus became "a symbol of innocence and purity."[20]

DesJardins mentions the writings of James Fenimore Cooper, Ralph Waldo Emerson, and David Thoreau as instrumental in communicating this Romantic myth of the wilderness to English-speaking inhabitants of the United States.[21] Even though the Romantic vision of the wholeness of the wilderness (including Muir's) inspires feelings of awe before powerful forces and wild savagery of nature, it is nevertheless an overly simplified vision. It depicts the wilderness as "pristine," as "unspoiled and uncorrupted . . . a symbol of innocence and purity" untouched by human interventions.[22] Hence the ethic of preservationism is for humans to preserve and leave nature alone, to leave it just as it is. Cittadino adds that the reception of this Romantic model of the environment was also promoted by "the promise of objectivity and a kind of moral purity in welcome contrast to the physical blight and moral decadence that characterized the urban and industrial environment of the late nineteenth and early twentieth centuries."[23]

Exhilarating though such a Romantic conversion may be, DesJardins cautions against this model. He says that Muir's defense of the wilderness "is based on the romantic wilderness model as described by Emerson and Thoreau. Because this romantic model of the wilderness underlies many environmental values, it is important to proceed cautiously. We need to be clear about how much this model is based on an accurate description of wilderness areas."[24] He points to three significant weaknesses in Romanticism's way of formulating the conversion to holism. First, he says that it perpetuates and exaggerates a false separation of human beings and their actions from the rest of nature. Second, he draws attention to the ways that Darwinian evolutionary thought constitutes yet another form of criticizing the nature/humanity dualism. Humans can hardly be regarded as separate from and parasitic upon nature if our species has also evolved out of "the wild forces of nature"—out of the natural "struggle for existence," as Darwin put it.

Third, this Romantic formulation of a pristine, natural holism is Eurocentric. It completely ignores the fact that Native Americans had been living, traveling, hunting, fishing, and farming in the allegedly untouched wilderness for thousands of years before European descendants began to move into these territories. The wilderness of this Romanticized formulation was not so much untouched by human beings. It was instead *continuously touched* by human beings in ways not readily noticeable to European newcomers. DesJardins quotes Chief Luther Standing Bear of the Ogallala Sioux: "We do not think of the great open plains, the beautiful rolling hills, and the winding streams with tangled growth as 'wild.' Only to the white man was nature a 'wilderness' and only to him was the land 'infested' with 'wild' animals and 'savage' people."[25] As DesJardins observes, "systematically ignoring or distorting" this fact "exhibits more than a small amount of cultural bias, if not outright racism."[26]

Conclusion

DesJardins presents the thought and activism of both Schweitzer and Muir as important breakthroughs (conversions) to a more adequate holistic foundation for environmental ethics. He also draws attention to the difficulties faced in trying to articulate the meaning of this holism. There is the internal challenge involved in attaining a sufficiently comprehensive and philosophically coherent meaning of the ideas of "value" or "the good" as a whole. There was also the need to get beyond the external cultural influences of the Romantic conception of the wilderness, which came to exert powerful influences in American thought in general. There was a third difficulty, already hinted at by the incompatibility between the scientific thought of Darwinian evolutionary biology on the one hand, and the tendency to separate a pure nature from human beings and their activities on the other. Early in the twentieth century the science of ecology would come into being, and its findings would pose even more far-reaching challenges for formulating the holistic foundation for environmental ethics. This is the subject of the next chapter.

Chapter 4

The Science of Ecology and Conversion to Environment as a Whole

There is a third important line of criticism regarding the Romantic symbol of the wilderness as a basis for ecological ethics. The symbol of a pristine, unchanging whole gradually became incompatible with the advances in the science of ecology. DesJardins and Cittadino present complementary accounts of the history of the emergence and development of the science of ecology and its impact on environmental thinking more generally. This chapter surveys their accounts.

The Development of Ecology as a Science

The term ecology was first coined by Ernst Haeckel in 1866 to denote the scientific "study of the relationships between organisms and what Charles Darwin loosely termed 'the conditions of life.'"[1] Yet "ecology" only began to be used in scientific literature toward the end of the nineteenth century. Ecology started as a subdivision of botany, which initially studied the morphology, physiology, and growth of isolated plants in laboratories. When botanists began to move out of laboratories and undertake numerous detailed field studies, they had to develop new methods. They needed new methods to study not just the structures of individual plants but also the interactions among plants. Field botanists had to study interactions among plants of different species, in widely different environments and during dramatic changes in weather.

The field studies originally focused on plant habitats alone. But the complex and dynamic natural environments soon required that ecology take account of ever wider ranges of conditions. First, animal ecology began to grow independently and parallel to plant ecology. This was followed by studies of the interactions among animal and plant populations. This required scientists to grapple with the problems synthesizing the methods and findings of both botanical and zoological ecological studies. Finally, ecological scientists had to also incorporate methods of the chemical and physical sciences, since air, temperature, sunlight, water, rocks, minerals, and soil affected how living organisms interact in their habitats.

These studies gradually revealed complex and dynamic patterns of change in natural environments. These discoveries forced scientists to abandon the steady-state Romantic model of nature in favor of various kinds of dynamic models.

The first model proposed to explain the dynamics of natural environments was called the "organic model." Different populations and species in an environment were likened to the different organs in a body. The changing interconnections among distinct populations and species were compared to bodily changes over the course of its growth and development. Thus natural environments were likened to developmental stages—nascent, infantile, adolescent, adult—as suggested by the organic analogy.

This organic model emerged out of numerous field studies of botanical environments. Frederic Clements, for example, studied prairies and grasslands in the Western plains. He "believed that for any given location, plant succession develops toward a stable [adult] and relatively permanent population [that] came to be called a climax community."[2] Some scientists such as Henry Cowles argued that environmental processes "are never complete" and do not have an adult or climax stage, but other field studies and theoretical arguments supported the climax community model for a while.[3]

Early in the twentieth century, however, the "organic model had begun to fall out of favor among ecologists. Natural biotic communities do not always develop toward some one single organic whole."[4] Cittadino narrates in detail the field studies and theoretical innovations that undermined the organic model. He also explains the scientific discoveries that showed that even modest changes in abiotic elements (e.g., water, soil chemicals, temperature) affect the dynamisms of biotic environments.[5] In the 1930s Arthur Tansley therefore introduced the term ecosystem to replace the organic model for environmental studies. For Tansley, "the whole *system* (in the sense of physics), including not only the organisms-complex, but also the whole complex of physical factors," need to be studied in their complex interactions.[6] The concept of ecosystem has been accepted ever since as the central scientific concept of ecology. Cittadino notes that Tansley used another idea, "Charles Elton's concepts of food webs and pyramids of numbers," in elaborating his concept of ecosystem.[7]

More recent developments in "complex dynamic systems theory" have provided further technical elaborations of the meaning of "ecosystem." "Elements within an ecosystem are related not simply in linear and causal ways, but also in more complex ways characterized as feedback loops."[8] The study of ecosystems focused on how nonlinear feedback loops operate among populations in nature.[9] Out of this research, two newer models emerged: (1) the community model and (2) the energy model.

(1) The community model analyzed ecosystems in terms of food exchanges. Entities in ecosystems played different and complex roles in the maintenance of the community, especially in the producing, distributing, and consuming of food. These roles were not restricted to the Darwinian model of food competition, since scientists discovered the importance of cooperative roles in food production in some cases.[10] (2) The energy model analyzed the dynamics of ecosystems in terms of changing patterns in which energy is exchanged among components in the system. This model gradually became the preferred scientific standard because it tended to bestow greater legitimacy upon ecology as a science. It deemphasized "such qualitative terms as *food, producers, consumers, communities and occupations* [replacing] them with seemingly more objective language of ecosystems and energy . . . [and] the mathematically more precise language of chemistry and physics."[11] It also incorporated the integral roles of abiotic elements—solar energy, water, and chemicals. The science of ecology eventually "focused on flows of energy and materials in complex multi-species systems."[12]

Need for a New Ethical Standard

Like the Romantic symbol of wilderness, the organic scientific model of environmental dynamics immediately suggested precepts for ethical response to the environment. The wilderness symbol had implied an ethic of nonintervention, but it ignored the dynamic realities of the environment. By way of contrast, the organic model suggested ethical ways of interacting with the environment that paralleled its way of understanding those dynamics. DesJardins notes that an ethics of "teleological reasoning would seem fitting" for the organic scientific model, since teleological reasoning "makes much of the connection between natural activities and the good." He continues: "From a natural science description of the normal development of the system (its equilibrium and stability), we could reason about what is good or bad, right or wrong, and healthy or unhealthy for elements of that system. Predators are good and ought to be protected, for example, because they contribute to stable populations within the system."[13] That is to say, in a teleological ethics derived from the organic scientific model the state of the climax community would supply an ethical standard for human interactions with any kind of environment. This would follow from the analogy of caring for the health of an individual organism, which would be a matter of acting appropriately at different stages in order to nurture its development toward the adult stage and thereafter supporting the mature functioning of that stage. By analogy, the organic model of natural environments implies that human actions ought to act in harmony with the developmental stages of environments so as to bring about climax communities, then to preserve them once they are formed, and to restore them from any damages they have suffered.

Unfortunately this teleological ethical approach was undercut when the organic model was replaced with the community and the energy exchange models of environmental dynamics. Neither of these models entails a climax community that could serve as a teleological standard for ethical actions. As DesJardins puts it, an ethical "ambiguity characterizes the implications drawn from a more chaotic view of ecosystems." When "ecologists reject the organic model, we seem to be faced with an even greater gap between ecological facts and environmental values."[14] If the dynamics of the ecosystems do not tend toward mature final stages, what ethical norms can we find?

Conversion to the Value of the Environment as a Whole: Aldo Leopold

One kind of response to this question comes out of a third conversion story that DesJardins incorporates into his history of environmental ethics: that of Aldo Leopold. Inspired by the example of Gifford Pinchot, Leopold studied forestry at Yale. He used what he learned there to develop a scientific approach to game management, initially adopting a utilitarian ethical framework. He applied his scientific training to maximize the "crop" of wild game (such as deer) for human consumption "by controlling the environmental factors which hold down" their numbers. Among the "controlling factors" were predators, especially wolves, threatening both wild deer and domestic livestock.[15]

Early in his career (1909), however, Leopold had an experience that would shatter his utilitarian outlook. In one of his most famous essays, "Thinking Like a Mountain," he proclaimed that "there lies a deeper meaning . . . in a hundred small events," especially the howl of a wolf. It is difficult, he continued, "to decipher the hidden meaning . . . only the mountain has lived long enough to listen to the howl of a wolf."[16] Inquiring into this hidden meaning, Leopold realized that wolves play important and integral roles in the *environment as a whole*. He tells of the life-changing moment that led to his conversion to the profound value of this wholeness, a narration that has become a classic among environmentalists. He had just shot but not yet killed a wolf. He described his experience as he approached the wolf to finish the task:

> We reached the old wolf in time to watch a fierce green fire dying in her eyes. I realized then, and have known ever since, that there was something new to me in those eyes—something known only to her and to the mountain. . . . I thought that because fewer wolves meant more deer, that no wolves would mean hunters' paradise. But after seeing the green fire die, I sensed that neither the wolf nor the mountain agreed with such a view.[17]

Leopold's conversion evolved, and he gradually came to understand that the "hidden meaning" was the value of the environment as a whole, with both its living and abiotic elements.[18] He went on to formulate a "land ethic" that came out of his conversion experience. His extensive field experience and research led him to understand the narrowness of the utilitarian conservationist approach to nature. Such an approach "seriously underestimates the interconnectedness of nature" and "treats the earth as 'dead' when in fact ecology recognizes that even a handful of dirt contains an abundance of living organisms."[19] As Leopold put it, "Philosophy, then, suggests one reason why we cannot destroy the earth with moral impunity; namely, that the 'dead' earth is an organism possessing a certain kind or degree of life, which we intuitively respect as such."[20]

It is noteworthy that Leopold says that we "intuitively respect" the earth. I would interpret his phrase "intuitively respect" as what Lonergan calls a feeling of intentional response to value.[21] Hence, to speak of the earth as having "life" is to use "life" as a symbol to convey the feeling for the value of the wholeness of the environment.

If one attempts to go beyond the symbolic meaning, however, the literal claim that the earth or land has "life" becomes both scientifically and philosophically problematic. Leopold was well aware of the limitations of taking the organic symbol too literally, so he qualified his claim with the words "possessing a *certain kind or degree* of life." He drew upon the findings of the science of ecology's energy model, writing, "Land, then, is not merely soil; it is a fountain of energy flowing through a circuit of soils, plants, and animals." It is this dynamic whole, this circuit, that is to be valued in a "land ethic." Leopold cast his new ethic in the form of a moral precept: "A thing [action] is right when it tends to preserve the integrity, stability, and beauty of the biotic community. It is wrong when it tends otherwise."[22] For Leopold, ethical evaluation of an action is not to be judged on the basis of its consequences for a particular *individual* organism. It is to be judged, rather, by its impact on a whole dynamic ecosystem. He drew upon and modified Tansley's ideas about ecosystems, incorporating them into the image of a pyramid.

The land pyramid is a "highly organized structure" of biotic and abiotic elements through which solar energy flows: "Each species, including ourselves, is a link in many chains. . . . The pyramid is a tangle of chains so complex as to seem disorderly, yet the stability of the system proves it to be a highly organized structure. Its functioning depends on the co-operation and competition of its diverse parts."[23] Ecosystems have a pyramidal shape because the numbers of prey must be larger than the predators on the next level. He regarded human beings as standing at the pinnacle of the pyramid. It is for this reason that Leopold believed he was justified to continue hunting throughout his life, notwithstanding the wolf experience of 1909.

Although Leopold's holism is much more sophisticated than those of his predecessors, there are still notable problems with his land ethic. Most of these problems derive from what scientists have come to understand about ecosystems—namely, that they are nonlinear, complex, dynamic systems that operate with feedback mechanisms. This poses a first difficulty because the results of any action upon such a system are inherently unpredictable. Hence it is not clear how any particular action whatsoever could ever be considered "right" in Leopold's sense. There is yet another difficulty. What would it mean to *preserve* a dynamic system, since it is always undergoing dynamic change? For certain mathematically modeled complex dynamic systems, preservation would simply mean not changing the differential or difference equations that specify the model. But almost no natural environments have been precisely modeled by means of such sets of equations. How, then, could one tell in advance (or even after the fact) whether a particular human action has changed the complex system itself? Notice Leopold's ambivalence regarding the words that function as symbolic expressions for the value of the whole: he uses "life," then "land," and finally "biotic community." This ambiguity in terminology betrays a residual organic metaphor and model latent in his thought.[24] Hence, Leopold's land ethic implicitly relies in part upon a stable system of nature, undercutting his own advances toward a more holistic, dynamic model.

Although Leopold's land ethic is more encompassing than those of his predecessors, his account of the whole from which environmental ethics must take its bearings remains philosophically unsatisfying. It leaves us with a significant challenge. As Cittadino puts it:

> [In the] late 1960s and early 1970s . . . the new environmental movement began drawing heavily upon the science of ecology for inspiration. . . . This was a role that some ecologists may have relished but for which most found themselves either unsuited or inadequately prepared. The science was itself incomplete, imprecise and conceptually inconsistent . . . ecology could offer limited assistance toward solving environmental problems, not only because of the enormous complexity of the problems themselves . . . but also because the solutions required complex value choices . . . that transcended the boundaries of their science.[25]

DesJardins adds: "The challenge for an ecocentric approach is to develop a coherent philosophical ethics that is consistent with ecology's emphasis on biotic wholes and yet recognizes that change is as normal as constancy. Thus we need an explanation of the change that governs this whole."[26]

We saw in chapter 1 that DesJardins raises the conundrum about the relationship between environmental science and ethics. He aptly criticizes the commonplace that science tells us what we should do. He draws important attention to Hume's criticism of the is/ought fallacy. But rather than showing how scientific knowledge can be properly related to ethical knowledge, he instead enumerates the several different traditions of ethical thought. At various points in his narrative, DesJardins comments on how one or another alternative ethical framework might be invoked to meet certain kinds of issues. Beginning in the 1930s, the science of ecology began to dramatically transform the scientific understanding of the environment, presenting an intensified need for some answer about the proper relationships between scientific and ethical knowledge. Aldo Leopold took a major step in this direction, but his thought still suffered from dialectical incoherencies that call for resolution. In chapter 7, I will explain further how Bernard Lonergan offered ways of addressing exactly these challenges. Before doing so, however, I will follow DesJardins' historical narrative to its completion in the next two chapters.

Chapter 5

Deep Ecology and Ecofeminism

Scientific studies increasingly revealed the damage human beings have been inflicting upon nature, things hidden from view of ordinary citizens. In 1964, biologist and talented writer Rachel Carson gathered up numerous isolated scientific publications and published a compelling synthesis of them to show the widespread and interconnected impacts of mass spraying of pesticides and herbicides. Once people began to understand the extent of human damage to the environment, they responded with indignation and sought the sources that led to such callous abuse of the environment.[1] Although Carson died of cancer a year and a half after publishing *Silent Spring*, she blazed a new trail for others to expose environmental damage and identify its sources. The next stages of DesJardins' history of environmental ethics—deep ecology and ecofeminism—take up the radical developments that arose in the wake of such revelations.

Deep Ecology

DesJardins opens his discussion of deep ecology with the following observation: "All of the environmental philosophies that we have examined so far can be classified as reformist. . . . This chapter will consider environmental philosophies that advocate more radical social change to address present environmental challenges."[2] The radical ethics of deep ecology arose twenty-five years after Aldo Leopold's death. *Deep ecology* is a term apparently first introduced by Norwegian philosopher Arne Naess in 1973. He was soon joined by sociologist Bill Devall and philosopher

George Sessions, the authors of *Deep Ecology: Living as if Nature Mattered.*[3] The term was soon adopted by such diverse groups of thinkers and activists that it became hard to discern any common definition. Naess and Devall therefore collaborated to formulate a manifesto or platform for deep ecology, which includes the following principles:

- The flourishing of human and nonhuman life has intrinsic value, independent of any utilitarian value for humans.
- Humans have no right to reduce the diversity of life except to satisfy vital needs.
- Human interference in the nonhuman world is excessive.
- Humans should promote the quality of life rather than a high standard of living.
- The flourishing of nonhuman life requires a decrease of the human population.[4]

There is quite a bit of overlap between these principles of deep ecology and the principles advocated earlier by biocentric and land ethics: rejections of anthropocentric, utilitarian, and individualistic ethical frameworks and the significant findings of ecological sciences. Yet deep ecologists regard themselves as radically different from their predecessors, whom they disparage as mere reformers. DesJardins explains that deep ecologists believe previous environmentalists were guilty of a serious oversight: "Deep ecologists trace the cause of many of our problems to the metaphysics presupposed by the dominant philosophy of modern industrial society. Deep ecology is concerned with a *metaphysical ecology* rather than a scientific one."[5] According to deep ecologists, modern industrial society is underpinned by "a metaphysics in which separate and isolated organisms are *most real.*"[6] This type of metaphysics emphasizes separation, especially the separation of humans from the rest of nature. Deep ecologists therefore claim that the "ecological and environmental devastation that has followed from this particular metaphysics has proved to be dangerous."[7] It has led to individualism, reductionism, objectivism, dualism (i.e., the radical bifurcation between humanity and nature), domination, capitalism, and even patriarchy (a critique that comes also from ecofeminist thinkers).

The urgency of our environmental crises therefore calls for a radical transformation of our metaphysical assumptions. The dominant metaphysics has minimized or ignored the reality of relationships. People must come to see that relations and processes are "at least as real, if not more real, than individual living organisms." In fact, "there are no individuals apart from or distinct from relationships within a system." DesJardins quotes biophysicist Harold Morowitz: "The reality of individuals is problematic because they do not exist per se but only as local perturbations in this universal energy flow."[8]

In this metaphysical shift from the primacy of individuals to the primacy of relationships, the "real world ceases to exist 'out there,' separate and apart from us."[9] Instead people should understand themselves as deeply interconnected with the rest of nature. Deep ecologists therefore contend that the distinction between objects and human subjects vanishes. In addition, this implies that the distinction between objectivity and subjectivity must be abandoned, since the notion of realities that exist independently of what humans think or believe must be eliminated.

This has also led to an ambivalent attitude toward the science of ecology on the part of deep ecologists, even though they continue to draw heavily from it. Naess, for instance, warns of the danger of an "ecologism" that would make the science of ecology ultimate. Ecologism would rely upon science to provide new technologies and "quick fixes" that fail to go to the root of the environmental crisis. Deep ecology therefore calls for something much more profound than new technologies. It calls for a new metaphysical framework within which ecological science is practiced.[10]

DesJardins draws attention to the performative contradiction into which deep ecologists have fallen.[11] On the one hand, their whole movement has been a radical criticism of the dominant metaphysics because of its destructiveness. On the other hand, their elimination of the distinction between objectivity and subjectivity leaves them with no basis for stating that their metaphysics "can claim privileged status as better reflecting reality." As DesJardins points out, deep ecologist are therefore "left without means for deciding which course of action . . . is more reasonable."[12]

In addition, there are sharp divisions among deep ecologists themselves over the proper ethical courses of action. Some advocate withdrawal into simple ways of living, others political activism, still others ecoterrorism. In general, though, deep ecologists do tend to agree in their opposition to human interference with nature: "Deep ecologists are committed to promoting lifestyles that tread lightly on the earth"; they are "less inclined to favor human interests" and even tend toward misanthropy.[13] This, again, is a performative contradiction. Theoretically, deep ecology is radically egalitarian across all life forms and vehemently opposes any form of hierarchical metaphysics. In practice, however, it tends toward a kind of inverted hierarchy: human beings tend to become the least important of all life forms.

In addition to his own criticisms of the performative contradictions of deep ecology, DesJardins also quotes Ramachandra Guha's charge that deep ecology is a culturally biased, "uniquely American ideology" that "smacks of Western imperialism." If put into practice, deep ecology would have "devastating consequences, especially for agrarian populations in underdeveloped countries."[14] New agricultural technologies (the "Green Revolution") of the 1960s saved over a billion people from starvation, especially in India. More recently, economic development has begun to raise billions of people in Africa and Asia out of extreme poverty. The policies recommended by deep ecology would roll these advances back and severely impact peoples in developing nations.

Deep ecologists were right to push the discussion to the metaphysical level. They appropriately criticized what Lonergan called the metaphysics of the "already out there now" reality. They were right to argue that relationships are not less real than individuals. Nevertheless, deep ecology was itself undermined by performative contradictions and the overgeneralizations exposed by Guha and ecofeminist criticisms.

Deep ecology has pushed the history of environmental ethics in an important direction. There is still a need for a metaphysics more adequate to all the complex realities of the environment. I will return to this issue again in chapter 7 and argue that Lonergan's metaphysics meets these and several other issues that have arisen thus far.

Ecofeminism

Ecofeminists have also criticized deep ecology for its abstractness and overgeneralization because it has ignored important distinctions among human beings. They disagree that a generic "humanity" can be held responsible for the domineering practices that have devasted the environment. Instead, it is the specific ideology of patriarchy that has fostered domination and oppression of both nature and women.[15]

The "logic of domination" is the key critique connecting feminism to environmental philosophy. Ecofeminists argue that the same spirit of patriarchal domination that has been responsible for the oppression of women underlies the ruthless exploitation of nature as well. The ways in which this connection is understood, however, depends upon how feminists think about the oppression of women. DesJardins draws upon the work of Val Plumwood and Karen Warren to describe "three waves" of feminism and the ways each wave has approached ecological ethics.[16]

Plumwood and Warren call the first wave "liberal feminism." It focused on fighting for equal rights, ending discrimination, and achieving equal opportunity for women, especially in education, all fields of employment and business leadership, political participation, religion, and social and family arrangements. Liberal feminists claim there is no relevant difference between women and men. They tend to hold "that all humans possess the same nature as free and rational beings and any unequal treatment of women would deny this moral equality and would therefore be unjust."[17] Plumwood and Warren argue that liberal feminism is uncritical, since women seeking to achieve equality in this way are in fact adapting themselves to an already existing patriarchal culture. Liberal feminism therefore expects "women to adopt the dominant male traits."[18]

The second wave is described as "radical feminism," which responded vigorously to liberal feminism. Some radical feminists saw the solution to the domination of women as the abolition of all sex and gender definitions and roles, but others advocated instead for positive valuation of what is distinctly feminine. "Accepting the view that some women do experience, understand, and value differently than men, some radical feminists seek to develop an alternative feminist politics, culture, and ethics."[19]

Plumwood explores how feminists in this wave contributed significantly to ecological thought in what she calls "cultural ecofeminism": "Rather than denying . . . the link between women and nature . . . cultural ecofeminists aim 'to remedy ecological and other problems through the creation of an alternative "women's culture" . . . based on revaluing, celebrating, and defending what patriarchy has devalued, including the feminine, non-human nature, the body, and the emotions.'"[20] Among other things, cultural feminists have developed a system of ethics based on "values important to women—caring, relationships, love, responsibility and trust" rather than "dominant modes in ethics" such as natural law, utilitarianism, and deontology, which "construe the moral realm in terms of abstract, rational, and universal principles."[21] Cultural ecofeminists have extended this approach into thinking about relationships that we have with, and caring for, the natural world. "More central for ethics are certain other questions. Do we care about animals? Do we have relationships with them? What is the basis for our attachment to animals?"[22]

The third wave of feminism arose in reaction to these trends in radical feminism, just as cultural feminism itself arose in response to liberal feminism. The third wave shares the conviction that "the domination of nature and the domination of women are inextricably connected." Third wave feminists, however, seek an alternative to both the first and the second wave positions. They regard cultural ecofeminism as still accepting the male/female, reason/emotion, mind/body dualisms that have been constructed by the patriarchy. Such dualisms "reinforce superior–inferior, oppressor–oppressed frameworks," so third wave feminists seek to expose them.[23] In particular, they have drawn attention to the metaphors used by Francis Bacon (among others), who held that men should dominate their wives and likened this to the attitude men should have toward nature. Bacon was the chief influence in the founding of the British Royal Society, so his way of linking the domination of nature to the domination of women had a great influence. Third wave ecofeminists have also explored alternative nondualistic ways of thinking about relationships to women and to nature that are "contextualist, pluralistic, inclusive and holistic." This approach is "holistic in that it encourages us to think about human beings as essentially a part of their human and natural communities."[24]

Conclusion

Both deep ecology and ecofeminism attempted to formulate more holistic ethical frameworks for approaching environmental challenges. They show the urgent need for a framework that can integrate the findings of environmental science while at the same time providing effective ways of making solid judgments of value about environments and ethical human relationships to them. Their own approaches subvert the possibility of arriving at the objective comparative judgments they need to effectively evaluate the environment as a whole and to criticize human actions. What is needed is a framework by means of which good answers can be proposed, debated, refined, accepted, and implemented.

According to DesJardins neither deep ecology nor ecofeminism has succeeded in finding the appropriate holistic framework that can address the issues raised by historical efforts to think about ethical responsibility toward the environment. He concludes this chapter of his narrative on deep ecology and ecofeminism by observing that their call for radical social change "has not fallen on fertile ground" and does not seem "to have had any lasting influence among environmentalists." He writes that more recent environmentalism "has taken on a pre-pragmatic shift" of "balancing environmental goods with the demands of economic and social justice."[25] We will return to this topic of a holistic framework again in chapter 7.

Chapter 6

Environmental Justice, Environmental Racism, and a Pragmatist Compromise

The concluding chapters of DesJardins' history of environmental ethics turn to issues of environmental justice, environmental racism, and a pragmatic approach. Because so many citizens of the United States tend to think of ethics in exclusively individualistic terms, DesJardins begins these chapters by emphasizing the social dimension of ethics. He writes that for ethics "the fundamental question [is]: How should we live?" He points out the often unnoticed collective dimension to the "we" of this question, which reveals that individual, personal morality is inextricably connected with social justice. The social dimension derives from the fact that we are recipients of benefits and burdens from others in society at large, and that our own actions contribute both to the benefits and burdens that impact others. Hence "How should we live?" inevitably includes questions of social responsibility for these benefits and burdens.

Environmental Justice

The question of social responsibility translates into the questions of the distributive justice—"How should the benefits and burdens of society be distributed? How should social institutions treat people? What do people deserve from society?"[1] DesJardins points out that ethics traditionally approached distributive justice within the wider context of "what is required to live a good, meaningful human life." Disagreements with traditional conceptions of the good life, however, have "led modern theories of justice to move away" from questions about the collective good life toward ideas of distributive justice based on the "rights of individuals to pursue their own conception of what is good." This has led to a prioritization of property: "Libertarian justice has long held that individual property rights and free markets are crucial elements of individual liberty."[2] Therefore the just distribution of benefits and harms in society came to be understood by libertarians overwhelmingly in terms of distribution and exchanges of property.

The libertarian approach to distributive justice traces back to John Locke. He claimed that human beings are originally and naturally in "a state of perfect freedom to order their actions, and dispose of their possessions and persons, as they think fit, within the bounds of the law of nature, without asking leave or depending upon the will of any other man." The "law of nature" however places some restrictions on this perfect freedom. According to Locke the law of nature prohibits each person from harming the "life, liberty, health or possessions" of any other person.[3] This law of nature ensures the equality of each person with every other person.

But what of human relationships to the natural world? Locke's inclusion of "possessions" in both the state of freedom and the law of nature elevate private property to an almost sacred status. This privileging of property skewed thinking about the natural environment in a momentous way. The natural environment, for Locke, is just "unowned," and it becomes privately owned "property" when an individual improves it by " 'mixing his labor' with the unowned land." The "person's exclusive rights over his or her labor are transferred to" the property.[4] Initially Locke seems to place some limits on this right to turn nature into one's own property, stating that one can use one's labor only to acquire as much property as one can use "before it spoils." For example, if one picks more fruit than one can eat before it spoils, this transgresses the natural limit to one's acquisition of property. Locke subtly dissolves most of that restriction, however, when he points out that "*a little piece of yellow metal* . . . would keep without wasting or decay." This unspoilable metal (money) can be received in exchange for the portions of one's property that cannot be consumed individually without spoilage. Since money cannot be spoiled, it can be kept as long as one likes and used in exchanges to acquire the property produced by the labor of others, including land or other natural resources others once cultivated. This effectively removes *all* limitations that Locke's principles of justice might otherwise have imposed upon what one can do to nature with one's labor. The notion that human beings might have *obligations* of any kind with respect to the environment (as opposed to other humans) is completely eliminated: "Numerous environmental concerns, from wilderness preservation to pollution controls to carbon emission regulations, and from wetlands protection to the Endangered Species Act, run afoul of the liberty and property rights of individuals. . . . Rights lie at the heart of [libertarian] social justice, and if environmental initiatives violate that right, they are unjust."[5]

DesJardins summarizes several objections that have been leveled against the absoluteness of this libertarian conception of property rights and the subordination of nature. He points out that the metaphor of "mixing" labor with elements of the natural environment can be taken neither literally nor in an unqualified sense. He further observes that the original human inhabitants of the North American continent had already been "mixing" their labor with nature in many ways long before the English-speaking settlers arrived. They did not, however, believe that their labor gave them exclusive rights of ownership and control.

DesJardins next considers the implications for environmental policies posed by an alternative theory of distributive justice, that of John Rawls in his *A Theory of Justice*. Rawls' theory incorporates some of the Lockean/libertarian principles, yet he sought to promote respect for all human beings, and especially their equal rights to liberty. This meant respect for each person's ability to make his or her own decisions, provided they do not harm or interfere with the rights of others to make their own decisions.[6] DesJardins explains how Rawls presented a nuanced and influential theory of "justice as fairness," which elaborates a basic framework for thinking about justice. He offers an imaginative exercise that supposes a person to not know his or her actual place in society. Such a person is then asked to think up rules according to which his or her society should be structured. Rawls argues that such a person would come up with two fundamental principles: First, "each individual is to have equal rights to the most extensive system of liberties." Second, "the social and economic benefits and burdens should be distributed equally unless an unequal distribution would benefit the least advantaged," but only if such distributions were accessible to all. In this way, Rawls provides both justification for the principles structuring a just society and broad guidelines for evaluating the fairness of social structures, institutions, and even individual decisions.

DesJardins offers only a very terse discussion the implications of Rawls' theory of justice for environmental ethics. He says that some have used Rawls' theory as a basis for the "precautionary principle [that] is often used in environmental policy making." According to this principle, if an action has the potential for great harm (for example, building an oil pipeline through a fragile habitat), very strong justification for that action must be provided.[7] Given the extensive influence of Rawls' theory, however, something beyond DesJardins' comments is called for.

Rawls' approach to justice poses a significant problem for environmental issues. First, his theory only addresses rights that human beings have toward one another within the framework of liberty and equality. Social arrangements are just if they meet the two principles regarding the distribution of goods and roles among human beings. However, as Michael Sandel points out, Rawls' attempt to respect the equality of all human individuals implicitly assumes an "unencumbered subject"[8] who would come up with and agree to those principles. That is to say, it treats all persons as equal to one another *at the cost of* pretending that persons do not actually exist in specific networks of relationships to other human beings. This also means that the human relationships to specific natural environments are also ignored in the development of these principles of justice.

Second, this approach focuses on the distribution of benefits and burdens among humans, so property rights again become focal. There is no room for evaluating how human actions bring benefits or burdens to nonhuman dimensions of the environment as such, only how the environmental impacts might visit benefits or burdens upon humans. Left in this state, Rawls' theory of justice is not particularly helpful for environmental ethics.

Following his discussion of Rawls' theory of justice, DesJardins raises a host of important questions of environmental justice: air and water pollution, toxic waste, wilderness and biodiversity loss, among others. He asks, "Who carries the burdens of environmental harms? Who would benefit from the policies promoted by environmentalists? Who would bear the burdens created by these policies?"[9] These of course are extremely important justice questions to ask. Yet de facto the answers have imposed the greatest proportions of the burdens on the poor while the wealthy reaped the greatest benefits. Rawls' work would seem to be relevant to these questions. Regretfully DesJardins does not turn to Rawls or any other philosophical theory of justice to respond to them. This is disappointing, especially in comparison to his own ongoing exploration of holistic approaches. A holism of the kind developed by Lonergan would be needed to provide guidance in the realm of environmental justice. I will return to this issue in chapter 7.

Environmental Racism

DesJardins reports that since the 1970s Richard D. Bullard and a number of other researchers and activists "have called attention to the disproportionate [environmental] risks faced by communities of color." These researchers have found that "toxic waste dumps, landfills, incinerators, and polluting industries" have been disproportionately located in neighborhoods inhabited by people of color and poor people. In 1982, a study by the United Church of Christ Commission on Racial Justice concluded that "race is the best predictor in identifying those communities and neighborhoods most likely to be the location of toxic waste sites."[10] The inhabitants of such communities tend to be overlooked by people with the power to make such decisions, and they often lack the economic resources and political connections to oppose such sitings. As a result, underprivileged communities tend to lose debates to those with more power to keep such facilities out of their neighborhoods. Bullard also demonstrates that governmental cleanups of environmental hazards and sanctions against polluters favored white communities and discriminated against poor communities and communities of color.[11]

More recently, researchers have also found that school, small business, and residential buildings located in minority communities suffer structural, respiratory, and other "sick" defects because they were built upon hidden waste dumps. In some cases, these buildings were built in locations that did not respect underlying geological features. For example, one school in Boston continually suffered cracks in walls and foundations because it was built over an old wetlands drainage pathway, as was later discovered. That pathway had been filled in over a century earlier, but the landfill did not entirely suppress the underground water flow. People who have the economic means to flee communities suffering such environmental disadvantages do so, leaving behind only those people who have no other choice than to remain.

DesJardins expands his discussion of environmental racism beyond the situations in the United States to the international arena. "Poor countries are more likely to suffer environmental degradation—deforestation, desertification, and air and water pollution—than wealthy countries, and the poorest resident of those countries, the poorest of the poor, are likely to suffer most."[12] He points out that for the most part this disparity is the legacy of colonialism, arising wherever Caucasian-European colonizers dominated nations inhabited by people of color. The colonial economic processes put into place had virtually no regard for the environmental consequences. Environmental questions seldom arose or were brushed aside.

Although justifications for overt colonialism are no longer acceptable, DesJardins points out that there are strong voices in economic theory and practice that continue to reinforce environmental injustices originally rooted in colonial practices. He quotes the highly influential economist Lawrence Summers, formerly director of the World Bank and US Secretary of the Treasury. Relying upon the economic norm of efficiency, Summers wrote, "From this point of view a given amount of health-impairing pollution should be done in the country with the lowest cost, which will be the country with the lowest wages. I think the economic logic behind dumping a load of toxic waste in the lowest-wage country is impeccable."[13] Not mentioned is that countries with the lowest wages are all nations populated by people of color, and that the "logic" is "impeccable" only insofar as it looks at global economic averages rather than national specifics—which just so happens to benefit Western countries where wages are higher.

In addition, proposals and policies that target population growth also tend to have racial and cultural biases to them. These policies especially target women in poor countries and tend to dismiss their cultural preferences for larger families. This is especially important in communities where proportionately fewer children survive to adulthood. He also quotes ecologist Garrett Hardin, who argued against famine relief programs because they would "only lead to a greater population explosion."[14] DesJardins points out the parallels to overt racist sterilization programs and writes that "minority people have reason to be skeptical of population policies advanced by wealthy white environmentalists."[15] While the rapid growth of the world's population does indeed have significant impacts on the environment and global warming, it is altogether too biased in favor of white people from affluent nations with low birth rates who would dictate policies for poorer communities. Other environmental policies have either neglected to address the disproportionate effects of environmental damage upon women or have actually had a greater negative impact upon women in poor nations. In chapter 19 we will consider a dramatically different proposal from Jeffery Sachs that addresses both issues of population growth and the impacts of environmental degradation upon women.

DesJardins also criticizes racial biases implicit in the ways that preservationists have affected the history of environmental actions. Policies to preserve forests and biodiversity can impede economic and social development in less developed nations. DesJardins writes, "It sounded as though northern environmentalists were saying, 'Our culture wreaked environmental havoc so that we might attain a comfortable and healthy lifestyle. Now that we have that, you should not seek a comparable standard of living, because that would jeopardize the remaining wilderness areas, rain forests and biological diversity.' "[16]

This wide range of injustices—national, international, explicit, implicit, historical, and present—is what is meant by environmental racism. Distributive justice means proportionate distribution of both benefits *and burdens*. It is therefore unjust to impose any burdens upon one group of people in disproportion to other groups without sufficient reason. Race, ethnicity, gender, or economic class do not constitute sufficient reasons.[17] Rawls' theory, says DesJardins, certainly would not accept these disproportionate distributions of burdens as just. He goes even further, writing, "It would be difficult to find any theory of justice that would accept any of these examples as just or fair."[18] Yet DesJardins himself has already explained how Lockean, utilitarian, and libertarian theories of justice rationalize such policies precisely as just and oppose environmental restrictions as unjust. Clearly a more holistic theory of justice is needed to account for environmental justice and racial justice together.

Since the publication of DesJardins' *Environmental Ethics*, there have been numerous books and studies published in the fields of both environmental justice and environmental racism.[19] It would take another book the length of this present one to do justice to all of this literature by applying Lonergan's method to their extensive research. However, since the plan here has been to follow DesJardins' narrative in order to demonstrate the relevance of Lonergan's method, I will have to limit myself to his account and leave a fuller engagement with more recent scholarship to another time. Hopefully this book will provide a model of how these issues can profit from Lonergan's contributions. I will, however, return to some of the issues DesJardins himself has raised about environmental justice and environmental racism in chapters 7 and 8.

A Pragmatist Compromise?

In his final chapter, DesJardins turns his attention to the problem of conflicts within the field of environmental ethics. He writes: "Irresolvable conflict about important matters does seem to threaten the foundations of an ethical life and our ability to know what is true" in general and about what is right in environmental matters in particular.[20] By "foundations" DesJardins is implicitly referring to a structure of ethical reasoning that spontaneously and implicitly operates in people prior to their engagement with explicit philosophical ethical theories. This is what Lonergan endeavored to access via his method of self-appropriation, and what he would propose as an approach to resolving such conflicts. We will return to this in the next chapter.

DesJardins offers his own view about the resolution that he thinks has been reached by the history of environmental ethics. Recent environmentalism, he writes, "has taken on a pre-pragmatic shift" of "balancing environmental goods with the demands of economic and social justice."[21] He argues that the conflicts he identified in his historical work do not justify extreme skepticism or relativism. Rather, he concludes his book with a proposal about pluralism and pragmatism: "It is time for philosophers to be more concerned with real-world practical issues such as pollution, environmental destruction, and environmental justice. . . . No single approach can be known to be correct in the abstract, apart from the particular context. Like Aristotle, pragmatists shift attention in ethics from what is *true* to what is *practical*."[22] Although he does not cite other contemporary thinkers in support of his claim about the current consensus, he does articulate a position that is heard frequently in present-day discussions of environmental ethics.

Contrary to DesJardins' exhortation, this is not a philosophically satisfying resolution. In fact, it is puzzling that he would cite Aristotle alongside the pragmatists in support of his position. It is true that Aristotle distinguished between the kind of precision appropriate to science versus the kind of knowledge appropriate to matters of practical reason. He also drew careful distinctions among theory, wisdom, science, craft (*technē*), and practical reason. Yet he never drew a distinction between what is practical and what is true. There is no denying, of course, that we are *truly* faced with extremely urgent environmental ethical problems. There is no denying that solutions to environmental and climate change problems must take into account very specific, practical issues that differ from one local situation to another. And there is no denying that much is already *truly* known about how to mitigate some of the most dire projected outcomes (although persuading the people, governments, organizations, and businesses to act along such paths is a greater challenge than the technical problems). There is thus a certain legitimacy to saying that people can act on what is now known, even if it will take much longer to sort out the conflicts about the proper formulation of a general environmental ethics.

DesJardins, however, claims more than this. He holds that the goal of reaching proper foundations for policies in environmental ethics should be abandoned altogether because the structure of human knowledge and the world make this impossible. As he puts it, "No single approach can be known to be correct in the abstract, apart from the particular context."[23] That itself, however, *is* a foundational claim, and for that reason it is what Lonergan calls a performative contradiction.[24] Philosophically, this cannot be accepted as the final word.

Neither, if Lonergan is correct, is this capitulation to pragmatism ultimately even a genuinely practical stance. As he put it, "Insight is the source not only of theoretical knowledge but also of all its practical applications, and indeed of all intelligent activity. . . . But to be practical is to do the intelligent thing, and to be unpractical is to keep blundering about. It follows that insight into both insight and oversight is the very key to practicality."[25] In other words, while DesJardins is certainly correct in saying that the proper foundations for environmental ethics do not lie in some single, abstract ethical theory, this does not preclude the possibility of an ethics based upon particular truths or "insight into both insight and oversight" (that is to say, self-appropriation).

It is at this point therefore that I transition from Lonergan's methodical specialties of History to those of Dialectic and Foundations.[26] Thus far I have been relying heavily upon DesJardins' and Cittadino's accounts of the historical movement of environmental ethics. DesJardins himself endeavored to move on from History toward something like Dialectic and Foundations in his concluding chapter, that is, to provide a way forward beyond the historically inherited conflicts that now permeate environmental ethics. He did so, however, without benefit of the methodical guidance that Lonergan offered for dealing with conflict—namely, the functional specialty of Dialectic. In the next two chapters we will consider how Lonergan's method can address some of the crucial issues that arose and remain unresolved in the history of environmental ethics.

Chapter 7

Toward the Wholeness of the Emerging Good

The preceding chapters were largely devoted to what Lonergan called the functional specialty of History. Lonergan did not intend to set out a new method or way of doing historical research. Rather, his method of eight functional specialties is intended to draw upon the work of scholars who use established critical historical methods and then to integrate their works into a larger framework that culminates in ethical policies and actions.

Accordingly, the preceding chapters relied primarily on the historical studies of DesJardins and Cittadino, which surfaced the challenges that have emerged during the history of environmental ethics. Relying on their studies, we saw: that utilitarian approaches to environmental ethics continually encountered difficulties; that the need for some kind of holistic ethical framework became ever more evident; and that while various thinkers attempted to formulate such frameworks, they nevertheless struggled to achieve coherent and philosophically sound holistic visions for environmental ethics.

This chapter presents a philosophical framework for thinking about environmental wholeness that meets many of those challenges. This framework is what I shall call "the emerging good." It is based upon Bernard Lonergan's notion of "emergent probability." Both Lonergan's notion of emergent probability and my extension of it into the notion of the emerging good are products of the functional specialty that Lonergan called Foundations. Foundations examines and formulates the meanings and implications of "conversions"—fundamental shifts in a person's assumptions about what is real, true, good, just, and ethical, along with corresponding shifts in assumptions about what can and cannot be known about such things. The conversions of Schweitzer, Muir, and Leopold have figured prominently in the history of environmental ethics and actions. Yet the impacts of their conversions were limited by the ways in which they formulated their meanings and implications.

Conversions happen with or without any reliance upon Lonergan's functional specialty of Foundations. That specialty provides resources to refine the interpretations of these kinds of conversions—in other words, to make those interpretations more methodical. Lonergan examined the dimensions and implications of these conversions by using his "basic method" of self-appropriation in order to connect conversions to the basic structures of knowing and deliberating that operate spontaneously in human consciousness. These connections enabled Lonergan to identify three dimensions or types of conversion, which he called intellectual, moral, and religious.[1]

The first sections of this chapter will summarize Lonergan's Foundational holistic notion of emergent probability. The later sections will show how his account of the emerging reality of the universe also implies a holistic ethical framework for thinking about the world—natural and human—as a dynamic, emerging good. In the next chapter and final chapter in part 1, I will offer a brief summary of the history of environmental ethics outlined in the preceding six chapters, using the Foundational notions from this chapter to highlight the Dialectical dimensions of that history.

Emergent Probability

Lonergan's notion of emergent probability envisions the wholeness of the universe and everything within it, not as a thing but as an evolving process. His way of thinking about the evolving universe takes recurring series of events ("schemes of recurrence" is his technical term), rather than things or species, as the fundamental building blocks of that evolutionary process. Once the most primitive schemes of recurrence begin to operate, they set the conditions for the emergence of subsequent and increasingly more complex schemes of recurrence.

The earliest schemes in emergent probability emerge according to the laws of physics. Later schemes subsequently and successively incorporate more primitive schemes as subcomponents and form increasingly more complex dynamic systems, including environmental ecosystems. These later schemes emerge and operate according to the laws of chemistry, biology, and animal psychology, in addition to the laws of physics. Still later schemes of human action, cooperation, organization, and civilization emerge as human beings have insights into possibilities and actualize them through their deliberations and decisions.

There is no necessity to the emergence of any particular scheme; schemes begin to function and continue to function only according to laws of probability. Once the earlier schemes start to function, this increases the probability of the emergence of later schemes. This is why Lonergan calls his notion "emergent probability." Emergent probability, therefore, characterizes the "upwardly but indeterminately directed dynamism" of the natural and human universe.[2]

Emergent probability is, therefore, the intelligibility of an unpredictable universe. Each newly emergent intelligible scheme contributes to this dynamism, but it is not necessarily a permanent, lasting contribution. Just as there is a probability of the emergence of schemes, so also is there a probability of their survival or demise. Emergent probability "neither denies nor minimizes such facts as entropy, cataclysm, the death that follows every birth, the extinction that threatens every survival. It offers no opinion on the ultimate fate of the universe. But it insists that the negative picture is not the whole picture."[3] The "whole picture" is neither any particular scheme or stage nor even a final end stage. The "whole picture" is rather the whole story of the emergent wholeness in which every emergent scheme and entity makes a contribution, however fleeting, to that story. This is the holism of Lonergan's notion of emergent probability.

Importantly, Lonergan arrived at his idea of emergent wholeness from his analysis of the methods of the modern empirical sciences. His philosophy of science is unlike that of virtually every other philosopher. According to Lonergan, scientists primarily seek to understand correctly the real intelligibilities of the natural and human worlds. Scientists begin, of course, with empirical data, but Lonergan presents a strong argument that more important still are the wonder and questions that scientists bring to those data. The scientist's questions seek insights into the intelligibility of the processes that produce those data. These preliminary insights provoke further questions about their own adequacy, correctness, and completeness. These further questions lead to further more refined and more correct insights. Science, then, is a self-correcting process consisting of a series of questions and insights such that insights suggest empirical implications, experiments and observations are undertaken on the basis of those insights, and experiments produce new empirical data that provoke further questions and therefore further insights. For Lonergan, science "is a circuit in which insights reveal their shortcomings by putting forth deeds or words or thoughts, and through that revelation prompt the further questions that lead to complementary insights."[4]

While most philosophies of science are preoccupied with how empirical data can prove whether scientific hypotheses and theories are true, Lonergan focuses instead on the growth in scientific understanding of the intelligibility of the universe. Truth of that understanding is not settled by primarily empirical data alone. Rather, the truth of hypothetical insights into intelligibility is settled both by empirical data and by the degree to which they answer all of the relevant questions that arise in a given scientific line of inquiry.[5] According to Lonergan, emergent probability is the intelligible wholeness of the universe that is gradually coming to be known through the ongoing efforts of scientists.

Scientists do not individually seek this intelligibility all at once. They seek the intelligibility of the universe communally, one insight and one verification at a time over the course of centuries. They seek each of those insights and verifications by utilizing methods that have likewise been developed over centuries. Lonergan therefore turned his attention to the methods scientists use in order to arrive at and revise their insights.[6] Although these methods are diverse, he discovered four general methodological patterns within those methods. It was on the basis of those four patterns that he developed his notion of emergent probability.

Lonergan devised his holistic notion of emergent probability on the basis of these methodological patterns, *not* on the basis of any particular scientific results of those methods. He endeavored to say only as much about emergent wholeness as could be said on the basis of the scientific methods rather than on their provisional results. In this way, his notion of wholeness was open to the dynamic, ongoing further developments in the sciences but not dependent upon any of their particular results.[7]

For this reason, Lonergan referred to emergent probability as a "heuristic notion." In his usage "heuristic notion" means "anticipatory." The most elemental form of a heuristic notion is a single question. Questions anticipate answers in the form of an insight, a judgment of fact or value, or a decision. At the same time, those questions set the standards as to which insights, judgments, or decisions will count as genuine answers. As both Lonergan and Aristotle observe, the "tension of inquiry" will not be resolved or go away for anything less than something that truly answers the question.[8]

While a single question is the most elementary instance of a heuristic notion, there are also more complex anticipations composed of complex interconnections among questions. Since some things can be known only in compounded sets of answers, Lonergan referred to these more complex anticipations as "heuristic structures." Through their sets of questions, heuristic structures anticipate a structured set of interconnected answers.

The notion of emergent probability is one such heuristic structure. It is an anticipation of the intelligible wholeness of the universe as it would be known if and when all scientific investigations manage to arrive at correct answers to countless numbers of scientific questions. As such, it is an anticipation of the intelligibility of the dynamic, evolving world as a whole. As a heuristic structure, it is not based upon the assumption of some grand theory or system but upon the structure of questions that anticipates an interrelated set of answers. Postmodern thinkers are justifiably suspicious of any pretense to have arrived at any *concept* of totality, and Lonergan was likewise chary of any such pretense. His notion of emergent probability is based instead upon scientific *questioning*, that is, the scientific seeking of correct understanding of the universe. Such questioning is radically open-ended. Hence, unlike the totalizing claims so severely criticized in postmodernity, the notion of emergent probability does not close off any questions. Nevertheless, its structured anticipations allow one to say *some limited* things about the intelligibility of the universe as a whole, without making absolute and finalized claims about all of the details. As Lonergan put it, emergent probability is about "the whole *in* knowledge, but not the whole *of* knowledge"; it points to "the general character of the concrete universe [but not] the concrete universe in all its details."[9]

The term *emergent probability* therefore has three closely related uses. First, it refers to a heuristic structure that anticipates and integrates all scientific methods, and at least implicitly, guides the whole enterprise of scientific investigations. Second, emergent probability refers to the actual intelligibility of the universe that is anticipated and sought by the heuristic structure of emergent probability. Third, it refers to the process governed by that intelligibility. Emergent probability, therefore, has a similar range of meanings as the more familiar term *evolution*. As a philosopher, Lonergan could assert a limited number of general things about the wholeness of that holistic intelligibility of emergent probability, based on his analysis of the methods of the sciences. But the ongoing process of the scientific enterprise gradually fills in the further concrete details of that wholeness.

The heuristic structure of emergent probability is itself a compound of two subsidiary heuristic structures—what Lonergan called classical and statistical heuristic methods. In spite of the great diversity of methods employed by scientists, Lonergan discerned that classical and statistical heuristics cut across these diverse fields.[10]

According to Lonergan, scientists utilize classical heuristic methods to discover "classical laws" that understand how events are "correlated" with one another.[11] Examples of classical laws are: Newton's law of gravitation, the gas laws of physics, Maxwell's laws of electromagnetism, the laws of chemical valence, Cuvier's law of correlation of anatomical structures in biology, and so on. Classical laws are formulations of the intelligible relationships among events in the universe. These classical laws or correlations explain the connections or relationships between one event and another: if A then B because of law_1; if B then C because of law_2; if C then D because of law_3, and so on.

Classical heuristic method is incorporated into the heuristic of emergent probability because of its use in the investigation of the recurring series of events (schemes of recurrence). Lonergan pointed out that in certain circumstances, sequences of events can circle around so that, at some point, event A comes about once again because of some other event in the series (e.g., if N then A because of law_n). If so, then one has not merely a sequence of events but a recurring scheme of events: A leads to B because of law_1; B leads to C because of law_2; C leads to D because of law_3 . . . N returns to A because of law_n, after which the cycle then repeats continually, "other things being equal."[12] Understanding the intelligibility of such schemes of recurrence is achieved through scientific investigations guided by classical method. Every transition in the scheme is explained by one of the classical laws, but the entire scheme is explained only by the whole structured combination of the laws.[13]

Classical method, therefore, is a crucial ingredient in the integral heuristic structure of emergent probability. The building blocks of emergent probability are the regularly recurring schemes in the universe. Classical method has led to scientific understanding of such schemes as planetary orbits, the circulation of water across the earth, the circulation of nutrients and water within plants, and the adenosine triphosphate (ATP) regeneration cycle that replenishes and supplies biologically useful energy to metabolic processes in certain animal cells. Classical investigations have also discovered that these schemes do not occur in isolation from one another. Most of these schemes occur within and are conditioned by ever larger and more complex schemes—ATP cycles depend upon Krebs cycles, which depend in turn upon mitochondrial schemes, and those are conditioned successively by other intracellular functionings, circulatory systems in animal organisms, and ecosystems within which animals are born, develop, live, and die. The entire science of ecology either used or discovered classical laws in order to understand and explain the complex, dynamically changing interdependencies among components within given environments.

The fact that schemes of recurrence function only under the proper conditions follows from the nature of classical laws of science themselves. In one of his most important contributions to philosophy of science, Lonergan pointed out that classical laws are abstract and therefore highly conditional.[14] That is to say, B will follow upon A according to law_1 only if certain conditions are fulfilled ("other things being equal"). For example, biochemical sequences will occur and recur only under proper conditions of temperature, concentrations, pH values, and so on. Classical laws by themselves do not dictate the conditions under which they actually operate. Classical laws do not necessitate anything. For this reason, schemes of recurrence will not emerge until appropriate conditions somehow come together.

Some conditions are fulfilled by other schemes of recurrence, but many times conditions are fulfilled at random. This provides the connection between classical and statistical scientific methods in the integral heuristic of emergent probability. As Lonergan put it, "Classical laws tell what would happen if conditions were fulfilled; statistical laws tell how often conditions are fulfilled."[15] Schemes therefore begin to recur (emerge) only when the appropriate conditions are in place. They likewise continue to recur only so long as those conditions endure. Neither classical laws nor their recurrent schemes fulfill all of their own conditions, so emergence and survival are matters of probabilities. In this way, statistical heuristic method adds a further complement to the classical heuristic structure. Together they are synthesized into the heuristic structure of emergent probability, which can guide toward understanding the complex processes of emergence, transformation, and demise that characterize natural environments. Classical and statistical methods in combination are therefore oriented toward and guide understanding of the emergent wholeness of the universe.

Finally, some schemes of recurrence are completely impossible under certain sets of conditions. There are no coldblooded reptiles at the South Pole. Once conditions begin to change favorably, however, some schemes that once had a zero probability of emerging now have at least a small positive probability. Such changes do not occur necessarily. They occur randomly, but nevertheless even random events occur in accord with probabilities (i.e., "ideal frequencies") discovered by statistical investigations. In many cases, the emergence of primitive schemes brings about a shift in the probabilities for the emergences of later schemes. This means that not only do new schemes emerge, but probabilities can also increase as new conditions come about. Along with the emergence of schemes, probabilities also emerge and increase.[16] This is why Lonergan called his heuristic "emergent *probability*." This is what gives the universe its "*upwardly* but indeterminately directed" character. On the other hand, while changing conditions as a whole conform to probabilities over long periods of time, individually they do so only randomly, which gives the universe its "upwardly but *indeterminately* directed" character.

Emergent Probability and Human Action

Human actions also occur within the overall holistic process of emergent probability. At the most elemental level, prehuman emergent probability brings about the conditions and environments to which human beings respond. It provides the environments in which human beings figure out how to survive and even thrive. Human beings participate in emergent probability by building upon the schemes of recurrence it provided before they arrived on the scene.

Human beings participate in emergent probability in ways that are markedly different from the ways that nonhumans participate. Nonhuman schemes of recurrence and their participants emerge, operate, survive, and perish variously under different sets of environmental conditions. They do so according to the various ways that the laws of physics, chemistry, botany, and animal psychology determine that they will under the various conditions. Nonhuman beings do not originate these laws nor how those laws determine emergences, responses, survival, or demise. Nonhumans function according to the ways these natural laws determine that they will function under given conditions.

By way of contrast, human beings respond to the conditions in which they find themselves by utilizing what I have called the structure of ethical intentionality. Human beings respond to their varying conditions first through their sensations, as do other animals. But unlike other animals, human beings spontaneously ask questions seeking understanding and knowledge of their circumstances. Based upon their answers to the questions about their situations, humans ask further questions and get insights into a number of different, possible courses of action that they could undertake in response to the conditions as they have come to understand them. These insights are like the laws of physics, chemistry, biology, animal psychology, which organize the events into recurring schemes. Human insights are unlike natural laws, however, because they are originated by the creativity of human intelligence; they are not determined by natural laws alone but build upon them.

Once they have arrived at insights into some possible courses of action, human beings next ask questions and deliberate about the outcomes and the values of these various courses of action. They can reach objective judgments of value, decisions, and actions as outcomes of these deliberations. Deliberations are guided both by the questions of value and by feelings that give humans an elementary consciousness (but not yet knowledge) of value.

Concretely most human insights and values implicitly or explicitly involve finding ways to cooperate with other humans in order to accomplish things of value. What might initially have been one-time cooperative ventures gradually become recurring patterns of cooperation that form skills, institutions, and social orders. In this way human cooperative schemes emerge out of and build upon the schemes of recurrence that constitute natural prehuman schemes of recurrence and ecosystems.

Ethical human action, therefore, is participation in and a prolongation of the intelligible process of emergent probability that has been going on in the world for billions of years. This is so, however, *only* insofar as human actions and cooperations result from truly objective deliberations—only insofar as deliberations address all relevant questions and are guided by fully converted horizons of feelings. But tragically so many human actions and cooperative ventures are taken within distorted horizons of value-feelings and before all questions have been faced intelligently and responsibly. Then the institutions and social orders are only partially intelligible and have mutilated values. Such has been the woeful legacy not only of how human beings have responded to one another, but also how human beings have responded to the conditions presented by the natural environments that were generated through the process of emergent probability. The lack of full understanding of natural ecosystems and the consequences that actions have, as well as truncated sensitivities to the full range of values and their proper scale, have left us with badly damaged environments.

Emergent Probability as Metaphysics

In chapter 5 we saw how deep ecology identified an erroneous metaphysics as the root cause of the environmental crisis. It called for a new metaphysics that would overcome the attitudes and behaviors responsible for environmental degradation. Lonergan used the framework of emergent probability to formulate his own metaphysics, and he did so in ways that meet the challenges posed by deep ecology without the drawbacks of their ways of formulating an alternative metaphysics.

Deep ecologists hold that in modern industrial society there is a "dominant scientific metaphysics in which separate and isolated organisms are most real."[17] They call instead for a radical transformation toward an "ecological metaphysics," which regards relationships as the most real, to the point of almost obliterating individuality as a viable metaphysical category. This also led them to claim that the "real world ceases to exist 'out there,' separate and apart from us,"[18] so that the distinction between objectivity and subjectivity must be abandoned.

Certainly relationships play a central role in Lonergan's heuristic structure of emergent probability. Relationships are what the classical heuristic method seeks to discover. Hence relationships are indeed very real components of the process of emergent probability. Lonergan makes this claim, however, not by abandoning science but by reinterpreting it through a new lens. For Lonergan, objectivity (scientific or otherwise) is not achieved by an isolated subject who separates itself from its object. That is an untenable notion of objectivity that assumes that separation is necessary in order to avoid projections and contaminations from the subject onto the object. According to this untenable notion,[19] objectivity would be possible only if a subject could know the object as it really is, isolated and "already out there now," apart from any subjective influences. Any prior knowledge relied upon by a subject as it attempted to know an object would destroy objectivity, according to the assumptions of deep ecology. Unhappily this would mean that no scientist could ever be objective if she or he were educated by taking science courses or learned from a laboratory director or from the findings of any predecessor—or even learned a language—since these influences would destroy scientific objectivity conceived of as absolute separation.

Like deep ecologists, Lonergan criticized the notion of reality as "already out there now"[20] and rejected it as an indefensible notion of reality. Unlike deep ecologists, however, he did not accept this as an accurate characterization of the metaphysics implicit in authentic practices of modern scientific methods. He argued, to the contrary, that closer examination (self-appropriation) of one's own knowing process reveals that objective knowing is achieved in the relatively small number of judgments that occur at the end of lengthy self-correcting processes of questions and answers. This is not a matter of separating oneself from one's prior insights, but of facing all the questions they provoke and seeking still further insights—further insights that would not have occurred without utilizing the prior insights. This self-correcting process, Lonergan argued, does not come about by separation but by engagement. Human knowing is not a matter of separation or confrontation but of self-transcending identification. Every question is a self-transcending moment, because questions draw us out of what we already have known toward what we do not yet know. Insofar as we answer questions with correct insights, our understanding draws us into the real intelligibility that we thereby correctly understand. This is the "position" on objectivity, because it is rooted in self-appropriation of what actually does occur in the consciousnesses of human beings as they engage in the process of knowing.

Insofar as scientists employ the classical heuristic structure to correctly understand the actual, complex relationships among things in the environment, they are subjects objectively knowing at least some features of the actual world that is distinct from themselves. But they are doing this neither by separating from those features, nor by obliterating the distinction between themselves and what they know. They achieve this by growing into understandings of the intelligibilities that constitute the realities they come to know.

Lonergan based his account of emergent probability on his cany analysis of the methods of modern empirical sciences. He later argued that this same heuristic structure could be established on the self-appropriation of the structures of human consciousness as such. In doing so he provided a foundation for emergent probability as a metaphysics that is more general than the methods of science that have been developed so far. He thereby argued that emergent probability is "the integral heuristic structure of proportionate being."[21] It is a metaphysics in which both individual things and relationships are coeval. Neither "isolated objects" nor relationships are privileged to the exclusion of the others. Relationships are between things. Things exist in relationships, and indeed in complex orders of relationships. Those orders are processes, series of events that are dynamically changing and forming an overall pattern Lonergan identified as emergent probability. Emergent probability is, therefore, an "ecological metaphysics" of the sort wished for by deep ecologists, without the performative contradictions pointed out in chapter 5. Things, relationships, events, orders, and processes are all realities that both scientists and ordinary people of common sense seek to understand as they really are.

The Emerging Good

The fact that human actions participate in emergent probability by building upon nonhuman processes has implications for the value and goodness of those nonhuman processes themselves. It has the implication that emergent probability has inherent value in itself, along with all the intelligible natural processes that it comprises. Lonergan provided an argument in support of this proposition.

Lonergan's heuristic notion of emergent probability is a structured anticipation of the emergent wholeness of the reality of the universe. As Hume rightly observed, however, from judgments of fact there can be no logical deduction of ought or any judgment of value or goodness. Still, human reasoning is limited neither to formal logic nor to matters of fact.[22] In response to matters of fact, people do raise questions of value and questions for deliberation and decisions about those matters of fact. Insofar as they pursue these questions in a serious way, seeking answers and refusing to settle for half-baked answers or biased substitutes, they are pursuing rational answers about matters of value and goodness.

This human reasoning from matters of fact toward objective judgments of ethical value and decisions has a structure, though not the structure of logic. Instead, it has what I have called "the structure of ethical intentionality." This structure of ethical intentionality has implications for environmental ethics. In particular, it is a structure of ethical reasoning that can begin with scientific knowledge of facts and proceed onward to sound ethical courses of action. It also has implications for the meaning of "the good." It especially has consequences for the senses in which animals, species, habitats, water, land, forests, and the environment in general all have inherent values and therefore have moral standing. Exploring these implications is the objective of this section.

Lonergan set forth his argument about the goodness of the universe of emergent probability in *Insight*. There he wrote, "The universal order which is emergent probability conditions and penetrates, corrects and develops every particular order. . . . [Accordingly] the realization of universal order is a true value."[23] I presented detailed analysis of his argument in *The Ethics of Discernment*. That level of detail is not possible here, so a brief summary will have to suffice.[24]

Any human decision—especially decisions to join in cooperative human actions—is a choice of something that is conditioned. Cutting down a tree is a choice that depends upon all the conditions that fostered the growth of the tree and made possible the cutting tools and skills. Starting a logging company depends not only on the conditions that led to the growth of a forest but also on all the conditions that make the skills and sustained strength of the loggers and the cooperation among them possible. A logging business also depends upon all the systems of transportation, marketing, and finance needed to sustain it as a recurring scheme. Lonergan therefore points out that it is abstract to think of either cutting a tree or operating a logging company as simple, isolated choices. If we move from abstractions to what is concretely being chosen in such cases, we realize that the object being chosen is something that is highly conditioned and, along with choosing it, one tacitly but no less actually chooses all the conditions that constitute it as a reality. As Lonergan put it, one "cannot consistently choose the conditioned and reject the conditions."[25] He further argues that the process of emergent probability is the grand process supplying all of the conditions through which all actual things and schemes of recurrence come to be. Emergent probability as the process that brings about all conditions is itself therefore the condition for everything and the condition of everything in the universe, past, present, and to come.

According to my analysis of the structure of ethical intentionality, good comes about when decisions are made on the basis of objective judgments of ethical value. Judgments of value, in turn, will be objective if all their "conditions are fulfilled." Fulfilling the conditions for an objective judgment of value is complex, but it includes asking and answering many questions about the accuracy of one's understanding of the situation, about the likely consequences of alternative courses of action, and about the normativity or distortions in one's feelings of value preferences.[26] In brief, objectivity of ethical decisions means taking into account (i.e., understanding and responsibly accepting) all of the conditions underlying the course of action one decides upon. Much of the understanding and taking responsibility for conditions and implications was already done tacitly during the long years of maturation where the consequences of one's actions were learned from family, neighbors, teachers, and others and how they reacted. Much of the learning to take conditions into account is also a matter of learning from one's mistakes, especially bad ones.[27]

Therefore, any choice of some concrete good or course of action, *if it were a fully objective choice of value*, would at least *implicitly* be choosing emergent probability because it is constitutive of that good in all its concreteness. Every human choice of some objective value therefore implicitly chooses emergent probability as good. Emergent probability is not a thing but a dynamic process, so good human choices implicitly affirm and tacitly choose the value of emergent probability that makes possible any concrete good chosen by a human being. I use the term *the emerging good* to denote the value of emergent probability.

This brief summary of Lonergan's argument no doubt sounds quite strange and perhaps is not persuasive.[28] This is because, as Lonergan observes, it "forces on [us] a new notion of the good . . . the good of order [that] consists in an intelligible pattern of relationships that condition" the fulfillment of our choices.[29] This means that what makes human social orders good is their intelligibility and value. This intelligibility is neither the ill-conceived intelligibility of half-baked ideas about how to organize people to get something done nor the fractured intelligibility of human social arrangements that ignore questions about consequences to others and to the environment. Human patterns of cooperation are good only insofar as they possess the well-rounded, corrected, and nuanced intelligibility and value that comes to be known in the self-correcting process of asking and answering the full range of questions regarding their implementations.

Since Lonergan described emergent probability as "the universal order which . . . conditions and penetrates, corrects and develops every particular order,"[30] it, too, falls under this new notion of the good as truly intelligible and valuable. This new notion of the good constitutes a break from the spontaneous identification of the good with "the object of desire" that underlies hedonistic utilitarianism.[31] This expansion of the notion of the good from objects of desire and pleasure to more holistic intelligible orders is what takes place in what Lonergan called moral conversion[32]—the kind of conversion experienced by Schweitzer, Muir, and Leopold.

Lonergan's argument for this conversion to a new notion of the good moves from egocentrism toward ecocentrism. It gradually entices the reader to broaden out to the full context of what really is at stake in any individual choice—nothing less than the environment as a whole.

The motto "Think globally, act locally" captures a parallel connection between the individual actions and universal stakes. The motto holds up the goodness of planet earth as a guiding principle for ethical action. One's local, individual, or group actions should foster and not harm that global goodness. Pictures of planet earth, especially those taken from outer space, convey feelings about the goodness of the planet and motivate people to think critically and act responsibly under the sway of those feelings.

Photos and other such symbols are exceptionally important in fostering feelings to guide environmental ethical actions. They can be misleading, however, because the photos of earth are unchanging while the earth itself and everything within it are dynamically changing through the process of emergent probability. The feelings evoked by such photos and symbols that evoke feelings for the values of the beautiful world therefore need to be supplemented by other symbols and expressions that will evoke feelings for the value of the dynamic process of emergent probability. These are needed for environmental ethics to reach full ethical maturity.

Of course, this argument for the emerging goodness of emerging probability rests on the assumption that there are human choices that actually are authentically good. It rests upon the assumption that the decisions people put into action are the result of having done the hard work of deliberating by asking and answering all the pertinent questions within the felt horizon oriented toward the wholeness of the emerging good. This includes but is not limited to asking and answering how the consequences of a particular decision will impact the very conditions that make human actions possible in the first place. It means not destroying the conditions, environmental and social, upon which human decisions build. To be fully concrete, it means not destroying the conditions that make possible the wholeness of the earth and its long-term future.

The Emerging Good and the Scale of Values

In general, human decisions are choices whose immediate contexts are human schemes of recurrence. Human choices are almost always choices to participate in, modify, or originate schemes of human cooperative action. Human beings come up with insights into how they can work together to accomplish some goals or purposes (values) while maintaining or improving their standards of living at the same time. The insights make the patterns of cooperation and their roles and institutions intelligible. The goals and purposes make them valuable—*if* they were arrived at through full and faithful exercise of the structure of ethical intentionality. Decisions to band together in intelligible and valuable ways produce genuine social justice.

Truly valuable and just human cooperative schemes are emergent upon prior, nonhuman environmental schemes. What Lonergan called the human good of order and value is a continuation of the emerging good of emergent probability and the environments it produced prior to the rise of humanity. Human cooperative social schemes modify at least to some extent the conditions under which the environmental schemes will function, just as the emergence of plant cycles modify atmospheric and water cycles, and the emergence of animal schemes modify plant cycles. *If* the human schemes are the results of authentic ethical responsibility, then the human schemes will be continuations of the movement of emerging good of emergent probability.

These observations connect with what Lonergan called the scale of value preference. According to Lonergan there are distinct kinds or levels of values in the emerging good. Because earlier levels of schemes of recurrence set the conditions for the emergence of later schemes of recurrence, there is a complex set of relationships between those that are conditioned and those that supply the conditions. At the prehuman level, there are distinct values corresponding to these conditioning/conditioned levels, which Lonergan identified as the physical, chemical, botanical, and sensitive animal (or "vital") orders of values. To these levels, human actions originate new schemes of recurrence, bringing about further distinctive levels of value, which Lonergan called social, cultural, personal, and religious. As he put it, "We may distinguish vital, social, cultural, personal, and religious values in an ascending order,"[33] to which should also be added physical, chemical, and botanical values as well.

There is of course intense skepticism about any claim of a hierarchy of values, including from advocates of deep ecology. A full explanation of Lonergan's argument for the scale of values and for its hierarchical ordering of values is not possible here. I have offered a detailed discussion elsewhere.[34] For the present, I will mention just two features of the argument for the scale of values.

First, Lonergan argued persuasively against "reductionism"—the assumption that every event in the universe can be completely explained by the science of physics alone. He argued, instead, that there are points in the process of emergent probability where a given science, say physics, is no longer capable of explaining all the regularities actually observed. At such points there arises a need, therefore, for autonomous sciences of chemistry, botany, animal psychology, and the human sciences. The sets of autonomous intelligibilities needed to fully comprehend the actual occurrences in the unfolding of emergent probability therefore implying distinct orders of beings—what he called "explanatory genera." The successive levels of values correspond to these levels of genera.[35]

Second, the tendency in philosophy and other disciplines to shun hierarchies certainly has its origins in abuses of assumptions about hierarchies both for the suppression of human freedom and for the exploitation of natural environments. For Lonergan, however, what is distinctive about the levels of social, cultural, and personal values is *not* that they make human beings so special that they can do whatever they please to everything "beneath" them. To the contrary, what constitutes these human levels as "higher" is the capacity and responsibility of humans to do what is right and of value for the whole of environments, not just what serves their own utilitarian desires. The proper response to the legitimate concern about hierarchies is not the pretense that there are not distinctive orders in the whole of reality or in the whole of the good. The proper response is, rather, critical understanding of these orders of value and their full ramifications. That is accomplished only by full and authentic exercise of the structure of ethical intentionality responding attentively, intelligently, reasonably, and responsibly to all the further questions that arise regarding the environment.

Conclusion

The heuristic of the emerging good answers DesJardins' call for a coherent philosophical ethics that is consistent with ecology's scientific understanding of the dynamism of environments.[36] I would state the heuristic guiding principle of this coherent philosophical ethics in the following terms: *Human decisions and the actions of human cooperative social institutions will be ethical if they continue the dynamism of the emerging good through intelligent insights put into action by objective judgments of value and decisions*. This ethical principle is heuristic. It is not a universal rule or a formula or a blueprint for the end to be achieved. Instead of exhorting people to follow some such blueprint, it exhorts them to discern and follow the spontaneous normativity of their own questions and self-correcting ethical intentionality. This guiding principle is grounded, as Lonergan put it, in the dynamism of emergent probability as it "becomes conscious." It becomes conscious in human questioning as it anticipates and guides the emergence of objective affirmation of the emerging good that is independent of human origination, along with the emergence of genuine goodness of human origin.[37] Human action is not ethical because it conforms to some predetermined plan. Individuals do not have to know the answers to every question about outcomes for the environmental and human future in order to be ethical. They do, however, have to commit themselves to discerning and explicitly following their own ongoing self-correcting process. This means paying attention to the questions one needs to answer before one acts on a given proposal. This includes turning to others who know answers that one does not know. Human actions will be ethical insofar as people commit themselves to continuation of the dynamic, emerging good of the world.

Wholehearted commitment to the wholeness of the emerging good and its full scale of values is what Lonergan meant by "moral conversion." This is a reorientation of one's feelings and decisions for values toward the wholeness of the good.[38] Such a reorientation is needed to sustain the pursuit of answers to *all* pertinent questions as the basis for ethical decisions in concrete circumstances. But moral conversion must also be complemented by intellectual conversion, which is open to all the questions about reality pursued by scientists as well as people of common sense. Intellectual conversion was the lens that enabled Lonergan to discern that the modern sciences seek to understand the intelligibility of the universe. Intellectual conversion creates dissatisfaction with the notion that science discovers only cold, brute facts.

Both moral and intellectual conversions need to be understood ever more adequately. As such they must be complemented by an adequate, explicit formulation of the heuristic of the emerging good if ethical cooperation with that emerging good is to be effective. That explicit formulation is the goal of the functional specialty of Foundations. In this chapter, I have been drawing upon Lonergan's own work in this methodical specialty, and adding a few points of my own from this functional specialty.

Of course, human actions all too often are not and have not been ethical. Questions that ought to have been faced were often ignored. Feelings that influenced ethical deliberations were often distorted. Actions were taken that disrupted and destroyed intelligible schemes of recurrence and values *without sufficient reason or compensating goodness*. Lonergan analyzed the various sources or biases that interfere with the self-correcting processes of cognitional structure and the structure of ethical intentionality.[39] These include the individual selfishness, group prejudice, and shortsightedness of common sense. Group bias in particular is manifest in the racism behind countless decisions that have placed the greatest burdens for environmental toxicity on people of color and others without economic or political power. These biases elevate some arbitrary, some unjustifiable factor such as greed, fear, race, economic status, self-satisfaction, favoritism, prejudice, or hatred of "the other" as the standard for decisions and actions. When this happens, these biases displace the desire to attain correct answers to one's own questions from its proper normative role in environmental ethics. When these biases interfere with the self-correcting process of human ethical intentionality and authenticity, they lead not to human cooperation with the emerging good but instead produce a "social surd" of social injustice where "intelligibility is only part of the whole."[40]

Human history is rife with examples of the biases that distort and divert the structure of ethical intentionality from its natural operation. These biases have led to injustices that have subverted the intelligible orders of social justice in human affairs as well as injustices toward the intelligible orders of ecosystems and the natural environment.

Nevertheless, human history is also filled with examples of people who worked against such biases to bring human actions back into alignment with the movement of the emerging good of the universe.[41] Lonergan studied such efforts and developed what he called the Dialectical method in the hope of making these efforts more effective. Much of the work of the thinkers narrated by DesJardins and Cittadino represents this kind of effort but without the benefit of Lonergan's methodological guidance. In the next chapter, therefore, I will show the advantages of bringing the Dialectical method into dialogue with the history of environmental ethics.

Chapter 8

The Dialectic of Environmental Ethics

From History to Dialectic

In the preceding chapters I have been summarizing the progression of environmental ethics as it unfolded in history. Beginning with Gifford Pinchot, DesJardins identifies the early adoption and eventual dominance of a utilitarian framework for approaching ethical and policy questions about the environment. He then traces the sequence of problems that the environment posed for utilitarianism along with various attempts to overcome those difficulties, including the revisions proposed in preference utilitarianism and extensionist utilitarianism. According to his narrative, none of these attempts were ultimately successful: "Environmental philosophers have reached a strong consensus that the narrow worldview of classical economics and preference utilitarianism that underlies it must be rejected."[1] He shows that the failures of utilitarianism to meet the challenges posed by the environment eventually led to radical breaks in the form of "conversions" to holism. Yet the attempts to formulate these conversions into ethical frameworks have been beset with difficulties and incoherencies.

This is the sort of historical process that Lonergan called "dialectical." That is to say, it is an oscillating series of historical events in which earlier positions and actions are met by opposing claims and responses, which are in turn met by still further oppositions. Lonergan's method draws upon the work of expert and dedicated historians to identify and narrate these dialectical movements in history. To their work, he adds the further contributions of a functional specialty that he called

Dialectic. The objective of Dialectic is to scrutinize the most fundamental disagreements and conflicts in history—the kinds of conflicts that cannot be resolved within the parameters of the contending frameworks themselves. These are the kinds of conflicts that he thought can only be resolved by some sort of "conversion." By "conversion" Lonergan did not mean conversion to a religion, a political party, a philosophical tradition, or a cause of some kind. He meant "conversion" to one's own true self.

Conversion to one's true self arises in response to strategic questions about oneself. Lonergan's method of Dialectic, therefore, depends upon taking seriously such questions as: "What am I doing when I am knowing?" "What am I doing when I am deliberating and acting ethically?" The method also follows up on the implications of answers to such questions.

Lonergan's method emphasizes what one is *doing* in the processes of knowing and deliberating. He argued persuasively that there is a spontaneous normativity inherent in those processes because they are spontaneously guided by living questions. Our spontaneous seeking of answers to our questions is normative only when it is followed without bias or interference. The method of Dialectic calls upon us to look past the *assumptions* we hold about the nature of knowledge and deliberation and to pay closer attention to the *operations* that *actually* take place in our processes of knowing and deliberating. We fall into a "performative contradiction" if our assumptions or claims about knowledge and deliberation are inconsistent with our actual operations of knowing and deliberating. Lonergan introduced the technical term "counterpositions" to characterize assumptions that contradict in language the normative performance of the operations of seeking genuine answers to our questions for understanding, knowledge, and value. On the other hand, claims about knowing and deliberating and valuing that are coherent with the actual normativity of our operations he called "positions." According to Lonergan, people committed to positions or counterpositions cannot persuade one another by logical arguments alone. This is because logical arguments always begin with undemonstrated premises; whereas positions and counterpositions *are the most basic* (often unacknowledged) undemonstrated premises held by their advocates. Only conversion to one's own spontaneous, normative processes of knowing and deliberating can settle these kinds of conflicts. Lonergan's "basic method" of self-appropriation (or what I have called the ethics of discernment) can address these deepest conflicts. His functional specialty of Foundations is designed to facilitate this self-appropriation, which can then be used in conjunction with Dialectic to provide a ground for resolving such conflicts.[2] In the preceding chapter I outlined how Lonergan himself answered the questions about knowing and deliberating, and how he spelled out their implications. I also included summaries of my own endeavors to extend Lonergan's work in the areas of ethical reasoning (deliberation), value, and the notion of the good that can provide more adequate foundations and guidance for environmental ethics.

In Lonergan's sense, counterpositions tend to be formulations of important ideas that posit as whole what in fact is only a part. Lonergan's stock example is the belief that knowing is identical with sensation ("taking a look"). He regarded as counterpositional the assumption that knowing is ultimately a matter of sensing, because sensation is not the whole of knowing. Rather, knowing is a self-correcting structure in which questions, insights, further questions, and judgments of fact add to the contributions from sensation to produce knowledge as a structured whole. Sensing is only a part of the whole of knowing, and its part is of relative, not ultimate, importance.[3]

Likewise, in the field of environmental ethics, utilitarianism and Romanticism constitute counterpositions, because they represent as whole what in fact is only part of the whole of value that is aimed at by the spontaneously normative operations of deliberating. Hedonistic utilitarianism is a counterposition because it assumes that the good is identical with the pleasant and evil is identical with pain. While authentic deliberating certainly takes pleasure and pain seriously, it goes beyond to ask whether there are any compensating higher values that could make some pains worthwhile, and to ask if there are higher values that might be lost in the pursuit of certain pleasures. The structure of deliberating therefore situates pleasure and pains as components within the larger whole of deliberating and the whole of value pursued by normative deliberating. Even revisions such as preference or extensionist utilitarianism represent as whole what are in fact parts within wholes. The former represents one's own individual preference as ultimate, when in fact deliberating spontaneously asks whether one's own preferences are correct, and how the preferences of others are to be taken into account. Still further questions ask whether even a majority's preferences might have blind spots and how minority preferences should be evaluated. Extensionist utilitarianism privileges the suffering or desires of individual sentient animals while overlooking questions about whether holistic phenomena such as biodiversity, land, or water that might be of greater value, or whether individual desires and sufferings are situated within a more holistic good. It is not individual preference, but the whole structure of deliberating and the emerging wholeness that it intends that corresponds to what is positional in ethics.

According to the narratives of DesJardins and Cittadino, the conversions to holism were important breakthroughs, but their full meanings and implications were not adequately formulated. Lonergan remarked that "every discovery is a significant contribution . . . [but if] it is formulated as a counterposition, it invites the exploration of its presuppositions and implications and it leads to its own reversal."[4] Adding Dialectical analyses to these historians' narratives, I would argue that this is exactly what happened as various figures attempted to express the meaning of a holistic foundation for environmental ethics. Lacking a full appropriation of the whole of value intended by normative deliberating, they formulated their breakthroughs in terms of counterpositions.

In particular the Romantic conception of nature tends to regard as whole what is only a part. It overemphasizes what Lonergan called the "vital level of value" as though it were the whole of value.[5] Hence, it fails to situate vital values into a more comprehensive, dynamically changing whole realm of emerging goodness. It is likely that most people in the US and elsewhere in the world today still tend to think about ethical responsibilities to the environment along the lines of the Romantic notion of wilderness, of not interfering with a mythical pristine nature.

The point to following DesJardins' narrative has been to show how the dialectic of environmental ethics has unfolded in actual history. Lonergan's method was invoked periodically to further illuminate deep sources of what has been going on in that historical process. Yet the tensions, conflicts, and conversions were taking place spontaneously without anyone's knowledge of Lonergan's ideas. They were spontaneously oriented toward a holistic resolution without Lonergan's guidance. Spontaneously they struggled with the formulations of that holism without reference to Lonergan. Nor did DesJardins rely upon Lonergan's method in crafting his historical narrative. He relied on the actual words of key actors, not Lonergan's, to identify "conversions" as key turning points.

What Lonergan's method adds is the identification of positions and counterpositions as deep sources of such conflicts. Positions and counterpositions refer back to conflicts within human beings themselves, conflicts that spontaneously call forth resolutions. Positions refer to the ethical norms that operate spontaneously as human beings deliberate about the right things to do. Counterpositions are the formulated assumptions that people rely upon in their ethical thinking, but are not coherent with their own spontaneous normativity, which they employ in their ethical deliberations. As Lonergan put it, "Any lack of coherence prompts the intelligent and reasonable inquirer to introduce coherence."[6] The thinkers DesJardins examines are unquestionably intelligent and reasonable inquirers, genuinely trying to make sense of human responsibilities toward the environment. The dialectical conflicts that DesJardins traces are ultimately between the spontaneous operations of consciousness that these thinkers employed in their thinking about those responsibilities versus the ways that they relied upon their explicitly formulated principles for such responsibilities.

The Dialectic of Environmental Ethical Thinking

The rest of this chapter repeats points made in earlier chapters but now makes explicit the dialectic of positions and counterpositions that was largely left implicit in the narratives of DesJardins and Cittadino.

This is especially true of the question of the relationship between scientific knowledge and ethical positions that runs through much of their narratives. In a sense, one could say that the dialectic of environmental ethics begins with Hume's declaration of the incommensurability between scientific knowledge of facts and ethical knowledge. In chapter 1 we saw how DesJardins spotlights Hume's argument that there is no logical deduction of "ought" from "is," He also reveals the continuing mischief that Hume's claim continues to cause to this day in reaching clear ethical judgments about environmental issues. He offers examples of how the same sets of scientific findings have been used to justify ethically opposed positions about environmental matters. He shows the difficulties that the ecological sciences encounter in providing philosophically acceptable grounds for deciding between opposing courses of action and environmental policies.[7]

This set of difficulties arises from Hume's counterposition about reasoning and knowing in general and ethical reasoning in particular. Hume's critique is effective only insofar as one concedes that reasoning can only be done in the mode of logical deduction. As the structure of ethical intentionality makes clear, however, the demands posed by questions for understanding and factual correctness, and the insights and judgments that respond to them, are more fundamental than the methods of formal logic.[8] In fact, the formulation of the methods of formal logic, which began in the work of Aristotle, are themselves the products of human inquiry, creative insights, intelligent formulation, and well-grounded judgments. While it would be irresponsible to ignore the norms of formal, deductive logic in their appropriate domain, logic alone is not the whole of human reasoning.

Ethical reasoning begins not with logic but with questions and answers about "what is going on." Important answers to these questions, especially regarding the environment, come out of the hard work and ongoing self-correcting efforts of scientists. Scientific answers provide the points of departure for subsequent ethical questions about what should be done in light of those answers. The processes of deliberation seeking answers to "What should be done?" inevitably lead to more questions of fact, some of which can only be answered by additional scientific research. Questions of fact, questions of practical possibilities, and questions of value interweave to structure a self-correcting process that is the basis of reasoning toward objective judgments of ethical value.[9] Therefore scientific, factual reasoning and ethical reasoning are not incommensurable with one another, as Hume's criticism alleges. Rather, the relationships among these complementary forms of reasoning are determined by the structures of inquiry in which both are embedded.

DesJardins also reports the debates that have arisen about responsibilities toward the future and especially toward future human populations. For example: "The 'argument from ignorance' stresses that we know little about people of the future. We do not know *who* they will be, *that* they will be, *what* they will be like, or what their needs, wants, or interests will be. Because we know so little about them, it makes little sense to try to specify any obligations to them that we might have."[10] It is not possible here to enter into a full discussion of the longstanding philosophical conundrum about whether we can know about the continued existence of the universe or anything within it. I offer, however, one point in response to that skepticism. It would restrict knowledge solely to knowledge of the present *because* it privileges sensation as the answer to "What are we doing when we are knowing?" Sensations are in fact confined to the punctual, immediate temporal present (unlike the contents of memories, anticipations, insights, or judgments), so it would seem that we would only be able to know the present for sure.

What we can and do know about the future, however, is known in the same way that we actually do know things about the present or past—not merely by sensing but by adding the further activities of inquiry, insight, reflection, and critically grounded judgment. DesJardins himself responds to the skeptical counterposition: "It is difficult to see why we should draw this conclusion. Surely we have a fairly good idea about what people of the future will need and what their interests will be if they are to have a reasonably good life."[11] While DesJardins does not explicitly subscribe to Lonergan's account of cognitional structure, he nonetheless is using that very structure in order to reverse the skeptical counterpositions about our responsibilities to the future of the natural and human environments.

A third issue involving a counterposition is whether individual animals, species, trees, water, land, air, or the earth as a whole have "moral standing" and "rights" that impose ethical obligations on humans. There is something both positional and counterpositional in the history of these debates. The notion of nonhumans having "rights" is positional because it already reflects a movement beyond restricting values and decisions to utilitarian human interests and concerns toward a more holistic realm of values. Such positions identify some values (expressed in terms of "rights") in the larger scale of values beyond the narrow range of human utilitarian values, which includes the vital values of sentient animals and the chemical and physical values of land and water schemes of recurrence. These discussions raise the further questions that must be properly answered about the values of human actions and the values (or harms) of the consequences of those actions for nonhuman entities.

Yet there is also something counterpositional, insofar as these debates base moral action on a notion of rights regarding what is "inherent in" agents. Such thinking tends in the direction of assuming that only what is "already out there now" in the agents endows them with "moral standing." The normativity of objective ethical intentionality does not look "out there" for something in some entity or in the environment in order to see if it has within it the "right stuff" to grant moral standing. Rather, the normativity of ethical intentionality is heuristic. It anticipates the values of entities and of courses of action as what is "to be known" as the outcome of authentic exercise of ethical intentionality. Knowledge of ethical responsibility arises as the end product of the exercise of self-correcting patterns of thinking that are faithful to the exigences of further pertinent questions about the values to be realized in actions. Ethical objectivity also arises out of a horizon of feelings that includes the whole range of the scale of values, from physical through animal to human. Ethical intentionality considers various courses of action by asking what values and harms would result and which values should prevail, and it comes to a rest only when all the pertinent value questions have been properly answered. These will inevitably include questions and judgments about the values of animal lives beyond their utilitarian values for human beings. When the outcomes of such deliberations are thorough and objective, they form the basis of objective knowledge of the values (or disvalues) of possible courses of action. Hence, a positional approach to the ethics of animal welfare is not based on rights or something like the capacity to feel pain, which somehow is supposed to endow animals with "moral standing." The positional approach focuses instead on self-appropriation of the process of one's own deliberating and decision making.

DesJardins' narrative also shows how the issues raised by human encounters with the environment inevitably forced people beyond the limitations of the utilitarian ethical precept: Act so as to maximize the sum total of pleasures minus pains. For Lonergan this is equivalent to confining the good to the lowest level in his account of the scale of human values, namely, the "particular goods" or vital values that satisfy individual human desires.[12] But when human beings act to achieve goods even on this level, they inescapably commit themselves to a far wider range of values. Almost any particular good that would satisfy some individual need can only be attained by participating in a complex human cooperative network of skills, cooperative efforts, and institutions that constitute what Lonergan called "the good of order" (i.e., social values). Goods of order, in turn, are sustained by cultural commitments, orientations, personal relations, and cultural values. These human cooperative complexes both condition and are conditioned by nonhuman ecosystems. Because, as Lonergan argued, one "cannot consistently choose the conditioned and reject the condition,"[13] this means that utilitarianism is actually committed to a much wider range of goods than it professes. It is tacitly committed to valuing all the schemes of recurrence and patterns of human cooperation that make possible the delivery of pleasures and avoidance of pains. It is also tacitly committed to valuing all the schemes of recurrence in nature that underlie those human schemes. Utilitarianism is a counterposition, for it fails to give an adequate account of what the utilitarian is actually doing even when he or she attempts to attain particular goods of pleasure and avoid particular evils of pain. Lonergan's remark about the counterposition of empiricism can be modified to read: "Utilitarianism amounts to the assumption that what is obvious about the good (pleasure) is what the good obviously is."[14]

Performative contradictions are unstable and call forth dialectical modifications as their assumptions encounter difficulties. DesJardins traces the dialectical series of attempts to modify the utilitarian framework in order to meet the ethical challenges posed by human encounters with the nonhuman environment. The first of these was the shift from "hedonistic utilitarianism" to "preference utilitarianism" in an attempt to overcome the insurmountable problems of quantifying pleasures and pains.[15] Preference utilitarianism substitutes satisfaction of what human beings *declare* as their preferred interests and concerns in place of quantifying and maximizing *actual* experiences of pleasures and minimizing *actual* experiences of pains. There is something positional in this modification, because human preferences are no longer limited to pleasures or pains, and actual human deliberation seeks comparative values beyond pleasure and pain. Gifford Pinchot is an interesting example of someone who used preference utilitarianism to formulate policies for himself and his agency, although it is clear from his actions that his actual range of values extended much further.[16]

Still, even the modification of preference utilitarianism does not provide an adequate account of or guidance for everything that people spontaneously believe to be good and evil in their interactions with the environment. Maximizing the satisfaction of human preferences and interests alone has led to consequences that people spontaneously feel and know to be wrong—the mistreatment of animals, for example. This led to a further modification of preference utilitarianism into animal welfare ethics, which attempted to redefine "interest" so as to include animals, but even this extension excluded too many values (and evils) that are held by most people who actually exercise their dynamic, self-correcting ethical intentionalities.

In spite of these difficulties, many environmental ethicists today still attempt to operate within the counterposition of utilitarianism and to find a version of it that will ground satisfactory ethical guidance. In fact, many environmentalists continue to rely on this framework at least implicitly in their attempts to gain acceptance for intelligent and responsible policies created to meet urgent environmental crises. This occurs when environmentalists try to persuade other people that it is in their own interests to preserve the environment. Most environmentalists who use this utilitarian language actually do believe that there are values at stake beyond human self-interests, but they adopt this rhetoric to persuade people stuck within a utilitarian mindset.

In DesJardins' account of the history of environmental ethics, this rejection of utilitarianism frequently came about through conversion to some form of holism. Such conversions are positional, for they recognize that there is far more to the good than can be accounted for by the utilitarian restriction to the level of particular goods. Each of the paradigmatic figures in his narrative—Schweitzer, Muir, and Leopold—nonetheless tended to *formulate* their conversion at least to some extent in counterpositional terms. Because of these counterpositional inadequacies, their formulations ran up against new kinds of difficulties. Among these were the ethical dilemmas regarding human "interference" with nature. Although Schweitzer refused to kill mosquitoes, he sanctioned other acts of killing for feeding or to end suffering. In addition, his biocentric focus overlooked the inherent values of nonliving entities such as water and land. Taylor attempted to modify and refine Schweitzer's principle of respect for life into a more nuanced biocentric theory in which "humans are seen as members of earth's community of life."[17] Unfortunately, his ethical principle of noninterference is inconsistent with the theory from which he derives it. It overlooks the fact that animals themselves already manipulate, control, and modify their environments by building nests, burrows, colonies, lodges, and dams, and by killing plants and other animals, among other things. If such animal actions are parts of life, how then can life be used as a principle to disallow similar human actions? DesJardins also cites a further difficulty: "It is not at all clear that species or ecosystems can be incorporated into the biocentric theory."[18]

Muir came to recognize the inherent value not only of all living things, but the whole of the wilderness itself. His writings and his untiring work to preserve wilderness led to the pragmatic outcomes of the founding of the US national parks, which have inspired tens of thousands of preservationists to this day. As DesJardins shows, though, the formulation of the value of wilderness as a whole became distorted by many of Muir's successors into the counterpositional Romantic myth of the wilderness, even if this is not what Muir intended. I would say that the counterposition in the Romantic myth of the wilderness consists in an elevation of vital values to an unwarranted priority in the normative scale of values that Lonergan identified. This elevation results in an unjustifiable demotion of the specifically human levels of value, namely, the social, cultural, and personal. This has tended in some cases toward misanthropy that regards human beings not as participants in the wholeness of nature but as alien parasites. The myth of a pristine, unchanging wilderness was also incapable of incorporating advances in environmental science. As the scientific understanding of the dynamisms of ecosystems began to abandon first the idea of the wilderness as unchanging and later the ideal of a climax community, the Romantic formulation of holism faced a dilemma. Insofar as it dogmatically ignored scientific findings in order to hold fast to its myth, it became enmeshed in both cognitional and ethical counterpositions. The alternative, as DesJardins notes, was a more radical formulation both of the wholeness of the good and the wholeness of reality that includes all that science comes to know about reality. In the preceding chapter I endeavored to show how the notions of emergent probability and the emerging good meet these difficulties on a positional basis.

In many ways, deep ecology also endeavored to move in this more radical direction. It called for an abandonment of a metaphysics that underpinned the unjustified exploitation of nature and its damaging consequences. In DesJardins' narrative, however, one can detect subtle but powerful remnants of the Romantic myth of the wilderness living on in deep ecology. As he points out, deep ecologists proclaim a metaphysics that is radically egalitarian across all life forms. They vehemently oppose any form of hierarchical metaphysics. In practice, though, they tend toward a kind of inverted hierarchy—human beings and their social structures tend to become the least important of all life forms. There is, of course, a distorted hierarchical metaphysics that has lent support to both damage of the nonhuman world and corruption of the human actors that are responsible for it, but the positional corrective to a corrupt hierarchical metaphysics is not no hierarchy at all. Rather, the corrective is a heuristic of the emerging good that replaces the corrupt hierarchy with an objectively normative hierarchy. In the scale of values as Lonergan conceived of it, the fact that human levels are higher than vital levels in no way justifies human beings doing whatever they please without proper regard to level of vital values. In fact, the reason the social, cultural, and personal levels are higher is because human beings are capable of bringing forth additional intelligibility and value to the intelligibility and value that is already emerging at the lower levels, including the valuing and protection of that nonhuman emerging good. The levels of social, cultural, and personal value are higher precisely because human beings can stop and ask about the actual intelligibility and value of the world and ask what further intelligibility and value can be built upon it and what aspects of it should be left alone. Doing whatever one pleases is no proper exercise of ethical intentionality.

In addition, DesJardins draws attention to how the egalitarian metaphysics of deep ecologists led to attempts to abolish the distinction between objectivity and subjectivity. Deep ecologists were responding to what Lonergan calls a counterposition on objectivity. According to that counterposition, human knowing and valuing is a confrontation between an objective reality that is "already out there now" and a human consciousness that is imprisoned in an "already in here now." Inevitably this idea about objectivity leads to alienation between human subjects and the harsh external "objective" world to which humans have no intrinsically valuable relationship. Lacking an alternative view of objectivity and subjectivity, deep ecologists thought they had no choice but to obliterate the distinction altogether. As DesJardins points out, though, this undermines their own claims about the superiority of their own metaphysics because it eliminates any objective criteria for the claim to superiority.

Yet there is a dramatic alternative to abolishing the distinction, an alternative Lonergan formulated as the position on objectivity: "Genuine objectivity is the fruit of authentic subjectivity . . . objectivity is the fruit of attentiveness, intelligence, reasonableness, and responsibility."[19] That is to say, objectivity about the right relationships and the right courses of action toward the environment is arrived at through judgments of fact and value that result from the sustained asking and answering of all pertinent questions within a normatively oriented scale of values.

Different from the deep ecologists who came somewhat after him, Aldo Leopold's formulation of his conversion as a "land ethic" overcame some of these problems. It was not limited to the values of individual organisms or even abiotic entities that must ethically be preserved. Rather, the "land community" as a whole was his foundational ethical value. His ethic took into account the earliest results from ecological science. Yet his formulation of his ethical precept of preserving the land community tacitly assumed some kind of climax community ecosystem as the standard for ethical action. The advances in the science of ecology have made this standard untenable and raised the need for an ethical framework that respects the actual dynamisms of ecosystems.

Moreover, Leopold's land ethic does not address the issues of environmental justice and racism. These are issues that concern the intelligible and properly valued integration of the interactions among human beings and between human beings and the environment. In other words, a fully objective environmental ethics asks and answers all the questions—not only how human actions impact nonhuman values but how consequences of those environmental impacts also affect human beings.

The holistic formulations of Schweitzer, Taylor, Muir, Leopold, and others are positional in their affirmation of the value of some whole that transcends individuals, whether humans or animals. They are also positional in their moral indignation toward widespread human disregard of the values of living and nonliving constituents of the world. Their formulations of their holistic conversions, however, tend to go too far in the direction of diminishing the values of human beings and the cooperative arrangements they have created. What is needed is a formulation of a holistic vision that meets all of the challenges that have arisen so far and is open to other challenges that lie ahead.

This is an all too brief summary of the dialectic of environmental ethics that DesJardins traces in careful detail. His work is impressive and positional in the careful and balanced interpretations that he gives to a long list of thinkers, activists, and writings. It is positional in the way he objectively presents the historical movement of environmental thinking away from utilitarianism and toward a more holistic basis for ethics. DesJardins has done impressive work as an historian, even though he is a philosopher and ethicist by profession.

There is, however, one place where, in my judgment, his narrative succumbs to a counterposition, and that is at its very end. There he writes: "Irresolvable conflict about important matters does seem to threaten the foundations of an ethical life and our ability to know what is true."[20] On the one hand, this has led to skepticism and relativism on the part of some philosophers and policymakers; on the other hand, it has led to outrage on the part of some environmental activists. Many activists disdain seemingly irresolvable philosophical debates when the need for action is urgent. In his concluding chapter, DesJardins rightly endeavors to reverse the counterpositions of skepticism and relativism that claim we really cannot know which opinions about environmental ethics are right. He likewise calls attention to the fact that we are running out of time to address human-caused climate change and that we cannot afford to wait until all the philosophical disputes are settled. He also points to the strong consensus that has emerged about concrete measures that need to be taken in order to address this crisis.

Yet in offering his response to this set of conflicts, I believe that DesJardins tacitly adopts a counterposition of his own. He urges the substitution of a pragmatic pluralism *over* truth.[21] Yet this is a performative contradiction on his part, for he is presenting pragmatic pluralism as a true and even obligatory, ethical way to proceed. Performative contradiction is always a sign (as Lonergan says) of a conflict between how one is *formulating* one's position and the *actual* ground of that position in the operations and structures of one's own consciousness.

What seems to underlie DesJardins' abandonment of truth as the criterion is his assumption that truth implies "embracing a single ethical theory."[22] He seems to be swayed by an underlying assumption that truth resides in a theory consisting of logical deductions from universally held principles to absolutely certain conclusions. DesJardins was influenced by William James and John Dewey in this regard.[23] The method of self-appropriation and the ethics of discernment that follows from it, however, are not "a single ethical theory" in this sense. The principles of an ethics of discernment are not a theory with propositions that function as logical premises with sweeping conclusions applicable to all circumstances indifferently. Rather, the *principles* of an ethics of discernment *are people* operating as best they can, employing their structures of ethical intentionality, and responding with fidelity to the promptings of their ever further ethical questions and the tensions in their horizons of feelings. When such promptings are taken seriously, they eventually call people to deepen their conversions to the whole of all that is good, to the wholeness of the emerging good. Hence there is no reason to oppose practicality to truth or to abandon one for the sake of the other. As Lonergan put it, "To be practical is to do the [truly] intelligent thing, and to be unpractical is to keep blundering about."[24]

Still, there is a positional kind of pragmatic pluralism that comes out of applying the structure of ethical intentionality in the vast array of diverse, concrete situations around the world. DesJardins remarks that even among people who have deep disagreements about their ethical theories, "a consensus is emerging around several ecological judgments" such as the limits of the natural environment and protection of air, water, soil, and access to food.[25] He describes situations in which "widespread agreement" has been achieved about concrete policies and courses of action even when partisans disagree strongly about ethical principles. He also tells of his own involvement in an environmental task force that agreed upon a number of concrete local actions in spite of their philosophical differences.[26] Groups of people do reach agreements about the courses of action that are truly valuable in spite of their theories. This, I would say, is the outcome of their exercises of their own self-correcting processes of ethical intentionality.

The exercises of ethical intentionality are indeed "pluralistic" in this very important way. The diverse, concrete situations across the globe in which people find themselves raise questions in one locale that are different from those in other locations. These questions call for appropriately different insights and valuable courses of action. Spontaneously people find fault with any ethical principle or universal bureaucratic policy imperative that fails to take into account such concrete differences and the specific questions they evoke. They find fault with any ethical theory that would suppress the natural movement of ethical intentionality as it raises all the unique questions pertinent to this or that concrete set of circumstances. To the extent that all the questions that are pertinent to a particular location are asked and answered, they will yield *true* judgments about the value of courses of action to be pursued. There are many different true courses of action insofar as they are appropriate to different concrete situations. Truth is not the special preserve of universal theories.

On the other hand, there is a counterpositional kind of pluralism that DesJardins falls into. It holds that truth is an obstacle to doing what is concretely ethical in concrete circumstances. It attempts to solve this problem by alleging that conflicting ethical theories can all be embraced simultaneously so long as they lead to the same concrete, practical outcomes. This convergence regarding practical outcomes may happen in a limited number of cases, but conflicting theoretical views will inevitably lead to opposed courses of action. The enemy is not truth but theories that do not recognize the normativity of a heuristic approach to the emerging good that respects all the relevant questions, practical as well as theoretical. A proper response to the set of conflicts presented by DesJardins comes from Lonergan's understanding of basic positions that arise from self-appropriation. The many conflicting environmental philosophies can be "contributions to a single but complex goal . . . enriched by adding the discoveries initially *expressed* as counterpositions,"[27] *provided* they are refined and transformed by an adequate method of Dialectic.

Conclusion

DesJardins draws three most important conclusions from his history of environmental ethics: (1) "Environmental philosophers have reached a strong consensus that the narrow worldview of classical economics and preference utilitarianism that underlies it must be rejected."[28] (2) "The challenge for an ecocentric approach is to develop a coherent philosophical ethics that is consistent with [the science of] ecology's emphasis on biotic wholes and yet recognizes that change is as normal as constancy."[29] (3) "The shift from individualism to holism within ecology is a helpful reminder that ethical issues arise both at the level of individual people and . . . must address questions of social justice as well as individual rights and duties."[30] His narrative points to the need for an ethical framework and a set of norms that can meet these three challenges.

It is here that the ethics of discernment and the structure of ethical intentionality have much to offer. This is because Lonergan envisioned generalized emergent probability and the emerging good as the intelligibility of the dynamic wholeness of the good of the universe. This has several consequences that address problems that have arisen in the history of environmental ethics.

First and most importantly, emergent probability is not merely compatible with the findings of the *present stage* of ecological science. It is instead based upon the scientific *methods* that heuristically anticipate the ongoing, self-correcting revisions of the sciences as they head toward ever more correct results. Emergent probability heuristically anticipates the ultimate sets of classical laws that will correctly explain the functioning of any actual schemes of recurrence. It envisions the ultimate, correct sets of probabilities that will explain the ideal frequencies according to which earlier schemes set the conditions for the emergence of later schemes, along with the probabilities of their demise. Finally, as was explained in chapter 7, the heuristic account of the reality of the dynamic universe set forth in emergent probability becomes the basis for a heuristic account of its emerging goodness. Emergent probability thereby provides a framework for an ethics that can direct efforts toward ethical responses to the environmental dynamics gradually discovered through scientific research.

Hence, Lonergan's holistic vision of the emerging good would substitute something in place of Leopold's ethical precept that "an action is right when it tends to preserve the integrity, stability, and beauty of the biotic community."[31] In its place Lonergan's ethical precept would be: "An action is right when it intelligently and ethically continues the process of emergent probability and the emerging good." Emergent probability is the process where even apart from human actions, new intelligible schemes of recurrence emerge in response to the opportunities provided by ever-changing sets of environmental conditions. Hence, to continue this in the human realm would be to act intelligently, reasonably, and responsibly in response to the conditions that are given in nonhuman as well as human environments. Unlike Leopold's tacit commitment to some climax community as the good to be preserved, generalized emergent probability "neither denies nor minimizes such facts as entropy, cataclysm, the death that follows every birth, the extinction that threatens every survival, [yet] it insists that the negative picture is not the whole picture."[32] The commitment to continue emergent probability confronts the ethical self "with a universe of being in which it finds itself, not the center of reference, but an object coordinated with other objects and, with them, subordinated to some destiny to be discovered or invented, approved or disdained, accepted or repudiated."[33] In other words, the environmental ethics of the emerging good is radically nonanthropocentric and is more profoundly ecocentric than even the most radical forms of deep ecology.

Second, emergent probability does not view the intelligent and responsible satisfaction of human needs, actions, or societies as unnatural or unethical impositions upon an otherwise good nature. Rather, it views them as components in the ongoing emerging good. The human good is one part of the ongoing emerging good. Just as the emerging human good builds upon the achievements of prior human efforts, so too it builds upon the goodness of abiotic schemes of recurrence and organic growth that precede, accompany, and condition all human achievements.

Third, the sequence of human actions, patterns of cooperation, and societies that constitute the human good *are good only* to the extent that they result from authentic exercises of the structure of human ethical intentionality. This means that decisions and actions have to ask and answer all the relevant questions about possible courses of action before they are undertaken, and this includes all the questions about impacts upon nonhuman as well as the human environments. Obviously, a great many human decisions have been made over the course of centuries that have ignored environmental questions that should have been faced but in fact were not. It is undeniable, therefore, that countless human actions have caused unethical harm to the environment. Nonetheless, this is not because humanity is a foreign intruder into or a parasite upon nature. It is, rather, because in failing to act according to the norms of the structure of ethical intentionality, human beings have forsaken collaboration with the all-embracing process of the emerging good.

Fourth, failures to act according to the norms of the structure of ethical intentionality have imposed injustices upon human beings as well as unjustified damage upon nature. As DesJardins shows, damage to environments has disproportionately impacted the poor, women, and people of color. These unjust distributions of burdens result because of deviations from the path of asking and answering all pertinent questions in concrete situations. Lonergan identifies "group bias" as one of the major interferences with the self-correcting dynamics of ethical intentionality. Group bias is the general structure of all the "isms," especially racism, sexism, and economic class preference. These biases all have their roots in distorted feelings within people's feelings for value preference. The biases operate when those in power allow the intersubjective identification with those like themselves to interfere with their self-correcting dynamisms of questioning and answering. As Lonergan wrote, "With remarkable acumen one solves one's own problems. With startling modesty one does not venture to raise the relevant further questions. Can one's solution be generalized? Is it compatible with the social order that exists? Is it compatible with any social order that proximately or even remotely is possible?"[34] Ethical intentionality and the heuristic of the emerging good direct human ethical choices toward actions that take into account all questions regarding proper and just care for fellow human beings as well as the natural environment, since both are intrinsic constituents of the emerging good.

Fifth, human pleasures and pains as well as the satisfaction of human wants, needs, and interests are all components within the structure of the human good and therefore components within the emerging good of the universe as a whole. Pursuits of human pleasure and avoidances of human pains are neither unnatural nor unethical as such. There is, however, a misleading tendency to regard all of these and the manners of satisfying them as somehow fixed. While there are "basic needs" for nourishment, protection, warmth, and procreation, their intensities as well as the manners in which they can be met are extremely variable. Such variability is a function of the irregularity of natural conditions, the unpredictability of human creativity (insight), and the diverse contemporary states of social and cultural schemes of recurrence.

Hence, while environmental ethics cannot ignore human needs and desires, neither can these provide the sole standard for determining ethical responsibilities toward the environment (as is the case in utilitarianism). This means, among other things, that all the pertinent questions about both human *and* animal pain must be taken into account in reaching ethical decisions about the environment. It does not mean, however, that pain is the trump card that decides the answers to all such questions. Animals inflict pain upon other animals as well as upon human beings, and this is part of the emerging goodness of generalized emergent probability. Paying attention to this fact and to the questions that emerge from it at least opens the possibility that in some instances humans would be justified in killing or inflicting pain on animals—provided that such actions are reached as the outcomes of authentic pursuits of the self-correcting process of ethical intentionality. It is the *whole* of the emerging good as continued in human ethical intentionality that provides the proper standard for answering such questions.

Finally, throughout the course of the history of environmental ethics, many people have been led to condemn anthropocentric approaches to ethics in favor of ecocentric approaches.[35] This is exactly what is achieved by taking emergent probability as the proper holistic account of the reality of environments and the emerging good as the standard for ethical action. Emergent probability is the ecosystem of all ecosystems. As Lonergan put it, intellectual development, especially science,

> reveals to [humanity] a universe of being in which [it] is but an item, and a universal order in which [its] desires and fears . . . , delight and anguish are but infinitesimal components in the history of [humankind]. It invites [humans] to become intelligent and reasonable not only in [their] knowing but also in [their] living, to guide [their] actions by referring them, not as an animal to a habitat, but as an intelligent being to the intelligible context of some universal order that is or is to be. Still, it is difficult for . . . the self as perceiving and feeling, as enjoying and suffering, functions as an animal in

> an environment, as a self-attached and self-interested center within its own narrow world of stimuli and responses. But the same self as inquiring and reflecting, as conceiving intelligently and judging reasonably, is carried by its own higher spontaneity to quite a different mode of operation with the opposite attributes of detachment and disinterestedness. It is confronted with a universe of being in which it finds itself, not the center of reference, but an object coordinated with other objects and, with them, subordinated to some destiny to be discovered or invented, approved or disdained, accepted or repudiated.[36]

Hence, the shift in the history of environmental ethics from an anthropocentric view toward an ecocentric view is implicitly positional, but this all depends upon how it is formulated. If it is formulated along the lines outlined in the previous chapter, it will tend toward being positional. But insofar as the formulation of an ecocentric ethics denies any kind of difference between human beings and nonhuman animals or asserts that the value of each human being is in no conceivable way higher than that of other creatures, then such formulations are counterpositional. For what is distinctive about human beings is that they *are drawn* to participate in the emerging good by their conscious tension toward all that is good, and that they *can* do so attentively, intelligently, reasonably, responsibly, and lovingly. While this distinctiveness cannot be used as a premise for permitting human beings to do whatever they please to other creatures, neither can this qualitative distinctiveness of the value of human beings be ignored without sinking into a counterposition and performative contraction.

This brings to conclusion our discussion of the history of environmental ethics, especially as told by DesJardins and Cittadino. In stark contrast to their impressive study of the history of environmental ethics stands their scant mention of the ethical issues arising out of global warming and climate change. Yet environmental issues are situated in the larger context of climate change. Environments are relatively local and are situated within the whole climate of the planet earth. The ethical issues regarding different kinds of environments, therefore, are likewise situated within the wholeness of global climate and global warming in particular. In part 2 of this book, therefore, we turn to these issues.

Part 2

The Ethics of Climate Change

Chapter 9

The Rise of Uniformitarianism

Introduction to the History of Climate Change Science

In part 1 of this book, I relied on the historical narratives of Joseph DesJardins and Eugene Cittadino to show the importance of the history of environmental science for environmental ethics. Part 2 continues that effort. However, the studies by DesJardins and Cittadino do not extend to the historical developments of climate change science or the changing ethical responses to it.[1] Part 2, therefore, will rely primarily on four works that narrate histories of climate change science and the cultural and ethical responses to this growing body of scientific knowledge: James Rodger Fleming's *Historical Perspectives on Climate Change*, Tobias Krüger's *Discovering the Ice Ages*, Spencer R. Weart's *The Discovery of Global Warming*, and Nathaniel Rich's *Losing Earth: A Recent History*.[2]

Before proceeding to the historical studies, there is need for a preliminary clarification of the term *climate change* itself. Climate change does not mean that it is hot one day and cold the next, nor that it is dry one day followed by precipitation the next. These are changes in the weather, not changes of the climate. The exact sequence of changes in the daily weather will differ from one year to the next, but the average weather patterns over many years constitutes the climate. Climate has more to do with the *pattern of averages* than the actual changes in weather that oscillate randomly around those averages. Those averages might remain constant in spite of the changes in actual weather from year to year.

Climate change, therefore, means that the averages themselves are changing. It means, for example, that the average temperature or the average level of precipitation rises continuously, not just from winter to summer in a given year, but across a succession of winters over a long period of time.

Global warming means the continual rise in the average temperatures over many years. Climate change means more than global warming. Climate change means changes in the averages of additional weather phenomena that follow upon global warming—such as continuous changes in average precipitation and storm patterns, in ice sheet coverage and ocean levels, in air and ocean currents, in areas of land that are arable, in frequencies of wildfires, and so on.

Finally, climate change has also come to abbreviate *anthropogenic climate change*, which is the claim that continuous changes in weather pattern averages are caused by human activities.

Almost two centuries ago scientists came to the conclusion that climate did *not* change at all, humanly caused or otherwise. Part 2 traces the history of how scientists first came to the position that climate did not change, then how scientists came to the realization that there had been ice ages (which meant climate does change). In their quest to understand what caused ice ages, scientists were gradually compelled to reverse their position, and proclaim that the planet is indeed warming and bringing about other correlated changes in climate. Finally, a scientific consensus that human activities are causing unprecedented levels of climate change emerged only in the last quarter of the twentieth century. The fact that a once well-established scientific theory had to be overcome contributed to the resistance in accepting anthropogenic climate change.

In part 1 we saw that environmental ethics became increasingly complicated as ecological science itself developed and offered an ever more complex picture of the natural world. The same is true but to a much greater degree when it comes to the science of climate change and ethical responses to its findings. Over the course of a century, scientific investigation inexorably produced an ever more complex picture of climate change dynamics. This in turn further complicates our understanding of natural ecosystems because climate sets the conditions underlying almost every other feature of the natural ecosystems discussed in part 1. The science of climate change pertains to the distribution of sunlight, heat, and water throughout the planet. The dynamic changes in those distributions throughout the earth are especially complex, and the development of a science adequate to understanding those complexities of climate took a long and winding course. As Spencer Weart puts it, "The tangled nature of climate research reflects nature itself. The earth's climate system is so irreducibly complicated that we will never grasp it completely, in the way that one might grasp a law of physics. These uncertainties infect the relationship between climate science and policy-making."[3] All the other complexities of life and ecosystems arise from and are reshaped, sometimes violently, by the dynamics of light, heat, and water. In part 1, we saw that the increasingly complex scientific understanding of ecosystems required a holistic ethics adequate to the emerging complexity of nature. Likewise, part 2 will show that the even greater complexity brought about by the climate dynamics underlying ecosystems poses further demands for our holistic ethics. One of the most important ethical challenges for the human race is thinking and acting in a way that meets challenges of such complexity.

Part of meeting this complexity is understanding the vacillating history of climate change science. James Fleming laments that the history of human thought about climate change "has not received adequate attention" and that climate change policies have not been sufficiently informed by it.[4] His narrative of that history focuses on the succession of "privileged positions" or "ways of making authoritative claims" regarding climate and how such positions came about.[5] Fleming's "privileged positions" roughly correspond to what Thomas Kuhn called "paradigms."[6] Fleming's notion of privileged positions emerged out of the questions that guided his study.

> I wish to understand how people became aware of climate change, how scientists and the general public understood the issues, how the study of the atmosphere changed over time, and the social and cultural implications of these changes. . . . How do people (scientists included) gain awareness and understanding of the phenomena that cover the entire globe, and that are constantly changing on time scales ranging from geological areas and centuries to decades, years, and seasons? . . . How were privileged positions created and defined?[7]

Fleming showed that the privileged position of human-caused climate change in the contemporary sense only began to guide serious scientific research in the late 1950s and only gained a wide scientific consensus in the early 1980s. This scientific knowledge was generated and accepted almost two hundred years after the Industrial Revolution began to dramatically increase the amounts of greenhouse gases (GHG) in the atmosphere. The phenomenon of human-caused climate change was happening for centuries before science began to comprehend it.

The details of the complex history of the science of climate change are presented in this and the next four chapters. It is a history that involves not only serial progress but also several dramatic shifts and reversals in the most fundamental commitments of the scientific community.

Many people, especially climate activists, claim that the facts about the human causes of climate change have been known for a long time, and that opponents are merely ignoring or obscuring what has been long known to be true. This shows up, for example, in Nathaniel Rich's claim, "It is incontrovertibly true that senior employees at . . . major oil and gas corporations knew about the dangers of climate change at least as early as the 1950s and did nothing to reduce emissions." Rich goes on to exaggerate his claim, writing that "everyone knew," including the automobile industry, utility companies, the US government, and environmentalists.[8] At best, Rich's claims and indictments could be justifiable after 1983 when the National Academy of Sciences issued its report *Changing Climate*. Even then, however, prominent scientists still had serious questions about climate predictions.[9] Weart presents an opposing view. He reports that the 1990 IPCC report

> predicted (correctly) that it would take another decade before scientists could say with any confidence whether the warming was caused by natural processes or by humanity's greenhouse gas emissions. Still, the panel thought it likely that human emission would cause global warming, perhaps a few degrees by the mid twenty-first century. . . . When the panel said the future was uncertain, that did not mean the risk to climate should be ignored. The experts had pointed out economically sound ways to get a start on reducing the risk, a broad hint that governments should begin to act.[10]

Furthermore, it was not until 1995 that the second IPCC report "agreed that human emissions might well be making the world warmer."[11]

In one sense, the idea that human activities can change the climate is very old indeed, but it is surprisingly an idea that was discredited by the advance of science itself. For that very reason, reestablishing the human causes of climate change faced formidable scientific challenges. The science that reestablished the connection between human activity and climate change had to follow a long and difficult scientific journey that lasted over a century.

Hence, when activists assume that the scientific facts of climate change have been obvious for a long time, they are overlooking the winding path that led to the current scientific consensus and how the complexities of that path have complicated the ethical challenges. The way forward involves convincing large numbers of people about the reality of human causes of climate change, about its impending consequences, and about the right things to be done in the face of this knowledge. In order to chart the best way forward, it is therefore essential to understand the natural history of climate change itself and the history of climate change science.

Early Cultural Sources of Thought on Climate

Fleming's narrative begins with the first of what he calls privileged positions, namely that climate *does* change and that human activity can bring it about. His narrative then traces how that position was eventually reversed by the mid-1800s through scientific research. The original position that climate did change was initially established, he says, through appeals to authority.[12] Originally this involved appeals to a literalist reading of the Bible, especially the Genesis accounts of the separation of the waters on the second day of creation (Gen. 1:6–10) and the story of Noah and the great deluge (Gen. 6:13–8:22). These biblical stories continued to have important influences into the middle of the nineteenth century, even influencing increasingly sophisticated attempts to provide scientific explanations for an ever-growing body of empirical observations.

Yet according to Fleming, an appeal to authority was also granted to Enlightenment figures as well as to religious traditions. Surprisingly, this path began with the widely held position in the eighteenth century that human beings *can and do* effect climate change, a view that grew out of the European settlement of the North American continent.

By the beginning of the eighteenth century numerous influential European thinkers (Du Bos, Montesquieu, Diderot, D'Alembert, and Hume) promoted a proto-theory of climate determinism that became widely accepted.[13] They held that the superiority of European to Native American culture was due to the more clement climate in Europe; they argued that the rough physical terrain and especially the climate of North America inhibits the development of human capacities for cultural advance and political self-government. They further argued that the cultivation of European lands had brought about this more temperate climate and worried that settlers in North America would devolve to the level of the indigenous peoples. "Many Europeans held considerable disdain for the New World and for its climate, soil, animals, and indigenous people."[14]

These negative views of the North American situation elicited patriotic responses on the part of the English-speaking settlers. "Colonials were quite defensive about these opinions and argued that the climate was improving as the forests were cleared."[15] Indeed it does seem apparent to common sense that the clearing of forests to cultivate crops would cause a rise in land temperature and a more moderate climate. The colonials favoring such a view included some highly influential figures: Benjamin Franklin claimed that winters were becoming less severe, and Thomas Jefferson even advanced a particularly elaborate theory that land clearance had changed the interactions of mountain airs and sea breezes.[16] English settlers applauded the value of these alleged changes as beneficial for meeting human needs and alleviating human poverty and suffering. Implicitly their defense and praise for human-caused climate change relied upon a utilitarian framework for evaluating what they thought were the benefits of human-generated climate change.

What went unnoticed and unacknowledged was that the alleged change was considered beneficial for addressing the needs of white, European, English-speaking people. They thought they were producing a climate "better suited to white settlers and less suited for natives."[17] The racial prejudices supporting these views blinded English settlers, although they are now blatantly obvious today. The views were also circular, since what defined a climate as "most suited to human habitation" was the one in England from which settlers came, a view that ignored the fact that the existing land and climate in North America had supported millions of non-European human beings for thousands of years.

In summary, this first privileged position drew on arguments from authority to claim that climate *does* change, whether by divine or human agency.

Data Collection as the Privileged Position

What was lacking in support of the first privileged position was widespread collection and systematic recording of data. Jefferson took one of the first steps to fill this lacuna. He "petitioned the U.S. Congress to include within the upcoming census considerations of the effects of the soil and climate on the population."[18] His concern was to study climate in order that human needs could be better satisfied. This utilitarian concern continued to motivate data collection in the century that followed. As Fleming observes of one major investigator, "Hemmer's stated motive was a 'precise understanding' of weather's influence on agriculture and health."[19] The motivation was *not* the holistic goal of better understanding climate as a reality and good in itself—a greater, encompassing good that needs to be taken into account along with human needs. Fueled by this utilitarian motivation, data collection gradually replaced human causation as the privileged position in this early stage of climate science. Even though the objective of these initial forays into climate science was learning how human efforts can change the climate in order to aid the improvement of human living conditions, they ironically led to the opposite, unforeseen conclusion, namely, that the North American climate was *not* being changed by human efforts.

At this stage, what counted as scientific knowledge of the climate was merely the collection of data. Fleming narrates the remarkable growth of meteorological observatories and societies in the United States, Europe, and Russia that began around 1819. This was a "tortuous and halting course of development" initially lacking accurate instruments and standardized procedures.[20] Eventually better instruments were constructed, standards for measurements devised, and observers were properly trained. Separate local data collections were shared and refined. They coalesced into coordinated national efforts that in turn grew into international cooperation. By the 1870s these developments produced the truly modern empirical sciences of weather and climate.

Fleming regards this period of the massive gathering of data as just one of the privileged positions. He narrates the correlative rise of scientific societies, university professorships, newsletters, journals, and standardizations that supported the ever more sophisticated gathering and recording of measurements. The growing influence of these scientific societies and practices effectively established data collection as the privileged position of climate science for a considerable period of time.[21]

Toward a Theorical Account of Unchanging Climate

Still, the massive amount of data collected over these decades lacked theoretical analysis and systematization. The first person to advance in this direction was Samuel Forry. Stimulated by Noah Webster's earlier negative criticisms of the quality of both the evidence and the arguments that had been advanced in support of human-caused climate change, Forry responded by comparing and statistically analyzing the data collected over the previous decades. On this basis, he argued that climate change had *not* occurred. In 1844 he published his findings, concluding that "climates are stable, and no accurate thermometric observations indicate systematic climatic change."[22] The renowned German scientist Alexander von Humbolt came to the same conclusion independently. The movement from data collection to theoretical synthesis that began with Forry and von Humbolt was advanced over the next four decades by a series of scientists.[23] They gathered additional data, refined the analysis, and strengthened the conclusion that there had been no substantial climate change in the United States, human caused or otherwise. As Cleveland Abbe put it in 1889, "No important climate change has yet been demonstrated since human history began."[24]

The Rise of the Science of Geology and the Uniformity of Climate

The many years of empirical observations and theoretical debates during the birth of the science of geology went even further in advancing the position that climate does not change. Fleming narrates how the science of geology began with the need to explain such phenomena as the stratification of different rock layers exposed in cliffs, canyon walls, and the rock surfaces exposed by recent construction of mines and canals. In 1799 a surveyor and engineer working for coal companies, William Smith, began to circulate an unpublished version of his findings that these sedimentary strata were continuous over vast geographical areas.[25] Early geological thinkers attempted to explain these layers by using modified variations of the Genesis story of the separation of the waters on the second day of Creation. They augmented the biblical story by arguing that this separation occurred over several stages, with Noah's flood as the last stage. Each stage of receding waters left a different layer of sediment that hardened into rock. These explanations were referred to as "catastrophism."[26]

James Hutton found these explanations unsatisfying. Between 1785 and 1795 he lectured on and published his alternative explanation. According to Hutton, the layering of different kinds of rock came about through alternating cycles. First came mountain building and uplift. Next, the uplifted rock materials were eroded and carried by streams and rivers into oceans where, over long periods of time, they formed layers of various kinds of sedimentary rocks. Eventually these layers were again raised above water level and once again eroded. This was a constant natural cycle. Hutton argued that this provided a better explanation than catastrophism for the geological phenomena known at the time. His explanation was also more in line with the Newtonian worldview of constantly recurring planetary cycles. It was the same process millennium after millennium. This also meant that the same unchanging climate—the same repeating cycles of weather—was responsible for the erosions. There were no catastrophic changes in climate. This theory came to be called uniformitarianism.[27]

The work of Charles Cuvier, however, raised difficult questions for Hutton's version of uniformitarian theory. Cuvier developed an impressive scientific method that showed how different parts of animal bodies were correlated with one another. His method made it possible to reconstruct the whole of an animal from fragments, sometimes even from a single bone. He applied this method to the study of fossils in different geological strata, establishing a correspondence between different types of animals and the geological strata of different ages. Once he established these correspondences, the different types of fossils could be used to date the rock strata. By reconstructing animals from fossil fragments, he was able to demonstrate that numerous animals that existed at the time of ancient geological strata could no longer be found in modern times.[28] While Cuvier himself did not resort to catastrophism and "special creations" to explain the extinctions and subsequent emergence of new life forms, others relied upon his work to do just that.

Partly in response to these claims by catastrophists, Charles Lyell developed a much more refined version of uniformitarian theory. Lyell was trained in law, and he knew how to marshal evidence in support of a claim. He assembled vast amounts of data, both mineral and fossil, and showed how the unchanging Huttonian cycles of uplift and erosion could explain them all. He also argued that what looked like worldwide extinctions and newly emergent life forms was based on highly localized evidence. He acknowledged that there could be "local catastrophes" but that there was insufficient evidence for global ones.[29] In addition, he claimed that the appearance of new fossil forms at different layers could be explained by migrations from other areas. Lyell had a gift for putting his explanations into a popular form, which he published in three volumes of *The Principles of Geology* between 1830 and 1833. Lyell expressed his overall objective of his book in its opening: "Historical sketch of the progress of geology, with a series of essays to show that the monuments of the ancient state of the earth and its inhabitants, which this science interprets, can only be understood by a previous acquaintance with terrestrial changes now in progress, both in the organic and inorganic worlds."[30] As Loren Eisley put it, "His book was widely read not only by professional geologists but by the cultivated public."[31] After Lyell's work, uniformitarianism became widely accepted as the best explanation ("privileged position") of geological and paleontological phenomena. It had a very profound effect on the development of Charles Darwin's formulation of his theory of evolution, and the eventual acceptance of Darwin's theory further enhanced the reputation of Lyell's work.

Lyell's Uniformitarianism also implied that there was no substantial change of climate over millions of years. Among other things, he investigated the fossilized records of rainfall, "the drops of which resembled in their average size those which now fall from the clouds." This meant "that the atmosphere of one of the remotest periods known in geology corresponded in density with that now investing the globe."[32] Lyell's work not only settled the controversy in favor of uniformitarianism in geological science but also implied overwhelmingly that there had not been climate change over the course of a million years of terrestrial history. By the middle of the nineteenth century, therefore, there was considerable scientific data and theory in support of the position that the climate does not change. This was the prevailing view among both scientists and the general public at this time.

Chapter 10

The Anomaly of Ice Ages

By the middle of the nineteenth century, the position of unchanging climate had been firmly established on many fronts by appeal both to evidence and to theoretical arguments. There were thus steep obstacles to surmount if the privileged position of unchanging climate was going to be reversed so that climate change could replace it as the privileged position. To do so would require not only new empirical evidence but also new theoretical explanations of the evidence, as well as shifts in cultural and political assumptions. These assumptions had been arrived at through extended critical debate and scientific research, so nothing less would be needed to overcome them. It would be wrong, therefore, to say people always knew they were imperiling the environment by industrial growth.

This chapter traces the rise of the acceptance of ice ages as real. This acceptance was the primary stimulus for the eventual overturn of the position that climate does not change. Here I rely on the historical study by Tobias Krüger, *Discovering the Ice Ages*.

While uniformitarianism had won wide acceptance by the middle of the nineteenth century as the privileged position in geological formation theory, there were problems on the horizon. Mountaineers and rural residents had long known about numerous gigantic boulders and massive chunks of rock that looked out of place and were often called "erratics" or *Findlinge*. By the beginning of the nineteenth century a great many of these objects were examined carefully and their differing stone characteristics were beginning to be cataloged and published. The reported observations of Genevan scientist Horace Bénédict de Saussure in particular were widely known.[1] The lithic characteristics of the erratic blocks were found to correspond to those of distant geological formations rather than to those of their current locations. Bavarian physician and astronomer Franz von Paula Gruithuisen, for example, argued that erratics found in the northern German lowlands originated in the mountains of Sweden over a thousand miles away.[2]

This naturally posed the question of how such massive rocks could have been transported over such great distances. Uniformitarianism alone did not explain the locations and distributions of these mammoth objects. Over the course of several decades, numerous explanations were offered. Among those holdouts who still accepted catastrophism, various flood mechanisms were proposed. As Krüger puts it, "During the first half of that century, a single overwhelming flood was becoming—to borrow the words of the German historian Michael Kempe—an 'increasingly swimming' notion. More and more scholars of the period regarded the scattered erratic blocks as relics of different floods, generally conceived."[3] Several versions of this line of theorizing involved water flowing down into large cavities and building up so much pressure that eruptions shot subterranean rocks out over large distances. Flood dynamics were also invoked to explain smoothing, rounding, and scratching of these rocks.

However, the extreme weights of these boulders posed serious problems for these explanations.[4] In response, mud deluges were also invoked.[5] One theory that gained considerable acceptance was that the gigantic rocks drifted upon ice floes until the ice melted and the rocks sank. This of course raised questions about the origins of these ice floes themselves. Rivers, tsunamis, or other no-longer-existing bodies of water were proposed as avenues upon which the ice floes drifted with their huge passengers.[6] While none of these explanations derived directly from the basic uniformitarian theses, some were at least compatible with them, functioning as a "protective belt" of auxiliary hypotheses, as Imre Lakatos put it.[7] Other hypotheses were not at all compatible with uniformitarianism, including flood explanations developed by catastrophists.

An alternative explanation involving glaciers began to emerge among scientists in the early part of the nineteenth century, although ordinary residents of the Swiss mountains had thought along similar lines a great many years. For example, a peasant told Johann Georg (Jean) de Charpentier in 1815 that "in former times the Valley of Bagne and Entremont had been entirely filled up by a glacier" and that this glacier had transported erratic blocks there. De Charpentier "completely rejected the idea at the time."[8] However, in 1818 a major disaster occurred with the rupture of the Glacier du Giétro that had been damming its lake. The resulting flood carried tons of debris and huge boulders, killing forty-four people. After this de Charpentier turned to the study of glaciers, although he still did not yet accept the idea of glacier transport of the erratics.

At about the same time other scientists had begun to associate erratics with glaciers. Several decades earlier, de Saussure examined the Mount Blanc region extensively and recognized that boulders had been transported by movements of the glaciers that were still present at that time. James Hutton read de Saussure's reports in *Voyages dans les Alps* and offered an explanation in the second volume of his own *Theory of the Earth*. Naturally, Hutton rejected catastrophist flood hypotheses and proposed instead that glaciers had transported the erratics to their present locations before the valleys were formed by erosion. "There would then have been immense valleys of ice sliding down in all directions towards the lower country and carrying large blocks of granite to a great distance." As Krüger says, "Hutton surmised the reason behind the greater glaciation was that the Alps had been higher than they are now. Being colder as a result, it would have favored increased accumulations of ice."[9] Presumably, the glaciers melted once erosion lowered the altitudes of the mountains. This was compatible with his uniformitarian theory.

Around 1815, Ignaz Venetz discovered that rocks falling into a glacier crevasse reappeared on the surface down the icefield. This was reported by de Charpentier at the Swiss Scientific Society's second annual meeting in 1816.[10] Venetz worked for many years gathering more observations in support of the former existence of a now vanished ancient glacier. As he continued his work, "his observations had led him to believe that not just Entremonts Valley but the whole of Valais had once been occupied by a glacier that had extended up to the Jura and had been the reason for the transport of erratic debris."[11] The idea of a glacier so vast was hard to believe. At about the same time Reinhard Bernhardi was also advancing the thesis that the polar ice cap had once been much larger, transporting the erratics to their present locations before retreating northward, but his thesis received virtually no support.[12]

In 1829, Ventz presented his hypothesis about the glacier, again at the Swiss Scientific Society, to an audience that included many prominent scientists. Most "did not consider Venetz's talk seriously," and the reception was very negative.[13] Initially de Charpentier himself "completely rejected this idea at the time" and thought that his friend's hypothesis was "really crazy and extravagant."[14] Out of friendship he decided to convince Venetz of the error of his ways. After many conversations and long hikes together inspecting the terrain, however, Venetz eventually persuaded de Charpentier, who soon became its strongest advocate in his lectures and publications. Along with Venetz and others, he began to gather more data in support of this hypothesis. In addition to the erratic blocks of rock, de Charpentier and others recognized that the same kinds of moraines (long ridges and piles of dirt and stone) that lay at the edges of existing glaciers could also be found at many other locations in the Alps where there were no glaciers. In 1841, on the basis of the distribution of erratic boulders and moraines, de Charpentier published an extensively documented map of a vast, now-vanished ancient glacier that once covered hundreds of square miles in the Alps.[15] After a period of skepticism, de Charpentier's relentless arguments eventually won acceptance by the scientific community as proof of the existence (and disappearance) of large-scale glaciers. Within a few years, others expanded the research and argued for the existence of other glaciers elsewhere in Europe and North America. These findings soon raised further questions about why the glaciers had disappeared and how they had originated in the first place.

The search for causes, however, ran up against a complication from another direction. Beginning around 1834 modern European thinkers began to speculate about how the earth itself was formed. First Emanuel Swedenborg and then Immanuel Kant and Pierre-Simon Laplace set forth increasingly sophisticated proposals about how the sun, earth, and solar system were formed out of hot gas that collapsed by gravitational force. This would mean that the nascent earth had been very hot and had been cooling ever since. Joseph Fourier knew that temperatures increase as one descends more deeply into the earth.[16] He conducted measurements and calculated heat flow from the interior to the surface, further supporting the thesis that the earth was continually cooling down. This thesis modified uniformitarianism, but it was accommodated to it with relative ease. The gradual cooling of the earth helped to explain uplift, deformations of the earth's crust, and mountain building. Even if the earth were gradually cooling, however, this did not seem to imply any change in the enduring climate patterns responsible for erosion.

De Charpentier's thesis raised more questions, though. If the earth originated as a hot mass, how was it possible that vast glaciers once existed and then melted, since the temperature was continually cooling rather than heating?[17] Krüger says that "de Charpentier therefore regarded the glaciation of the alpine region as well as of Scandinavia as a temporary, local phenomenon."[18] De Charpentier had encountered problematic data, but he had not yet abandoned uniformitarianism, and a global theory of climate change was still years away.

Karl Friedrich Schimper was among those immersed in studying the connections between glaciers and erratics. Initially he subscribed to the idea that erratics were deposited from ice floes and that the growth and decrease of the glaciers could be explained by a combination of mountain building and erosion due to their successive impacts on cooling and warming. But he added something new: that the whole earth went through several cycles of cooling and warming with successive growths and recessions of glaciers. In between, a "generally temperate climate" had prevailed. However, he initially thought that the temperate period was ending and that "cold is on the advance."[19]

In July of 1836, Schimper met his friend from university days, Louis Agassiz, along with de Charpentier at the annual meeting of the Swiss Scientific Society. Shortly thereafter, Schimper did more exploring of the Swiss Alps and read de Charpentier's article on "the probable cause" of the movement of erratic boulders in the Swiss Alps along with his hypothesis of the super-glacier.[20] A few months later Schimper and Agassiz stayed in Devens, Switzerland, where de Charpentier had a villa, and they were occasionally joined also by Venetz to discuss erratics and boulders. After they left, Schimper continued to make new discoveries about glacial debris, including the fact that their distribution was not random, so they could not have merely fallen from ice floes. At about the same time Agassiz also conducted more research, concluding among other things that the smoothness of rock surfaces came from glacial polishing, not water erosion, as had been previously thought.[21]

Schimper then spent the winter at Agassiz' home as his guest, making additional observations in the nearby mountains and engaging in many long conversations. Their conversations led to the eventual formation of what Krüger calls "The Grand Synthesis" of the ice age theory by Schimper and Agassiz. It "offered the key to explaining numerous phenomena for which science had hitherto been able to offer only partly satisfactory hypotheses."[22] Together they "channeled influences from various contemporary scientists, ranging from Cuvier and Bernhardi to de Charpentier and Venetz."[23]

Krüger summarizes the key elements of the synthesis:

- A significantly larger extension of alpine glaciers across the Swiss Central Plain up to the Jura Mountains in prehistoric times.
- The Alps rose up only after the ice sheet had formed. When the new mountain range broke through these masses of ice . . . erratic blocks and debris had rolled down the ice.
- A sudden and puzzling drop in the ambient temperature, which extinguished all life on earth. This temperature drop was part of a cyclical pattern characterizing the entire history of the earth and had a lasting influence on the evolution of life.
- Ice once covered not only Switzerland but also large parts of Europe.[24]

Schimper and Agassiz each began to present these ideas separately in a series of talks. Then, in July of 1837, Agassiz presented his ideas at the annual meeting of the Swiss Scientific Society. Krüger describes the content of the speech at length:

> [Agassiz] claimed that the glaciers now visible in Switzerland had once been substantially larger. Conceding outright that he was borrowing findings by Venetz and de Charpentier, Agassiz then reported about observations he had made in the Jura [Mountains] from which he deduced that this range too had once been enveloped in ice. Finally, he went a step further and postulated that the whole of Europe had once been covered by giant glaciers, from the North Pole up to the Mediterranean. The same situation existed in prehistoric North America and Asia. Presupposing that the Earth had originally been very hot, it had cooled down a little more after each new geological epoch. The climate during each individual epoch had been similarly stable to the present day. However, each epoch had ended with a temperature drop that had "produced freezing cold weather." Afterwards the

planet had warmed up again but without attaining the same temperature level as the preceding epoch.[25]

Krüger narrates that "reaction by the audience ranged from laughter to indignation," that even de Charpentier "was shocked about this wild hypothesis," and that the decorum of the lecture hall deteriorated—all in spite of the fact that Agassiz was then president of the society![26] The next day Agassiz read a letter to the society from Schimper providing numerous empirical details in support of the theory, but the members remained unconvinced.

Matters deteriorated further when periodicals began to cite Agassiz as the discoverer of ice ages, with no mention of Schimper. Among many other things, it was Schimper who coined the term "ice age."[27] In spite of Schimper's request to have his contributions acknowledged, Agassiz continually, even deliberately, failed to give him due credit.[28] Later on, Agassiz also slighted the contributions of others as well, including de Charpentier. Not until twenty years after Schimper's death were his true contributions and Agassiz' shameful treatment of him made known to scientific societies.[29]

Nevertheless, the ice age hypothesis eventually prevailed. Krüger provides a long and detailed account of the difficult, separate pathways to acceptance of the ice age theory among scientists in each of several different nations (France, Great Britain, Sweden, Finland, Russia, Germany, Canada, and the United States) between 1837 and the mid-1860s.[30] Krüger writes, "It was only when this hypothesis of one or more ice ages began to extricate itself from its catastrophistic context that it became acceptable enough for gradualistically arguing geologists in the 1850s and 1860s."[31] The final triumph came, perhaps, with the publication of James Croll's monumental work *Climate and Time* in 1875. By then, "one no longer needed to be a catastrophist in order to acknowledge glacial periods as fact."[32]

Krüger's point about the break with catastrophism is important, but it needs a balancing emphasis. While some of the strongest objections to the acceptance of ice ages as a fact came from those who were suspicious of them as a new form of catastrophism, the theory of ice ages nonetheless posed a serious problem for strict uniformitarianism. If there were ice ages, climate patterns were not always the same. The reality of ice ages posed what Thomas Kuhn called an "anomaly" for the uniformitarian paradigm (or privileged position) that reigned in the geological and climate sciences during the middle of the nineteenth century. To be even more precise, gigantic erratic rocks were the anomalies that led to the further anomaly of ice ages. This fact cried out for some explanation. Searches for this explanation led in the direction that would eventually become the science of climate change.

Krüger's thorough research into the rise and acceptance of multiple ice ages illustrates several features of Lonergan's account of the growth of science as described in earlier chapters. It was not merely data but the questions they raised that led to the insights about glaciers as potential causes, and then to the insights about large, no longer existing glaciers. The scientific way of proceeding is to wonder about data, get insights, reflect on the insights in order to see under what conditions they could be confirmed, and then to seek more data, raising more questions and leading to more insights that adjust or abandon earlier insights. Questions and insights drive the search for further data. Scientific growth stems from the self-correcting cycle until sufficient volumes of data and insights are assembled and questions are answered to form something approximating a virtually unconditioned, verified set of judgments.[33]

Although the self-correcting pattern permeates the growth of the science documented by Krüger, he also shows that there were biases operating in the scientific community that were not open to further questions. He shows, for example, that Agassiz impacted the movement of science not only through his careful attention to data and his brilliant insights but also through his selfish, individual bias, which not only prevented proper credit to Schimper but delayed the insights by other scientists that would have come had Schimper's insights been recognized. Krüger shows that acceptance of scientific hypotheses and theories involves more than just observing sense data. Acceptance comes through having questions awakened and being persuaded that the new insights do answer those further questions. He also shows that a complex and initially diverse set of data elicits a complex set of questions and requires a complex set of insights, and how insights from diverse scientific investigations had to be brought together, thereby revealing the need for something like the integral heuristic structure discussed in chapter 7. Moreover, biases had to be overcome eventually, in order for the science of ice ages to become widely accepted.

To summarize: through the combined and dialectical efforts of scientists such as Venetz, de Charpentier, Schimper, and Agassiz, ice ages were accepted as explanatory and historical fact, and uniformitarianism was overthrown. This path was complex and winding, but the growth of the science of climate change itself would prove even more complex and dialectical. This will be the topic of the next three chapters.

Chapter 11

Joseph Fourier and the Science of Heat Dynamics

In his history of the science of climate change, Fleming highlights the importance of the theoretical contributions by French mathematician and physicist Jean-Baptiste-Joseph Fourier. Fourier is often credited as the first person to advance a scientific theory of human-caused climate change. Although Fourier's memoir of 1827 is frequently cited as the first scientific statement of the greenhouse effect, Fleming goes to great lengths to show that this is an exaggeration. His criticisms help to sort out just how much Fourier had worked out and how much had to wait upon the work of his successors.

The path began with Fourier's seminal work on the dynamics of heat, *The Analytic Theory of Heat* of 1822. In its introduction, Fourier expressed admiration for Isaac Newton's achievement because his "small number of laws" explained the most diverse phenomena: "the movements of the stars, the inequalities of their courses, the equilibrium and the oscillations of the seas," among other things. Fourier aspired to be "the Newton of heat" and to construct a "cosmology of heat" to rival Newton's cosmology of the planetary system.[1] It was his greatest ambition to give a complete theoretical account of the dynamics of heat distribution throughout the whole earth. He observed that in spite of the great advances of Newton and his successors, none of their accomplishments could "apply to the effects of heat."[2] Fourier therefore formulated a new physical law of heat movement and developed powerful new mathematical methods for applying this law to a wide range of situations and diverse conditions.

Some substances absorb more heat than others—that is, they require more heat to raise their temperature one degree than is required by other substances. Likewise, heat flows faster in some substances than others, and it flows faster where the temperature differences are greater. Fourier found an elegant way of integrating three phenomena in one single law covering the conditions that govern how heat is absorbed by a substance, how heat flows through the substance, and how heat is radiated or reflected from one substance into another substance. On this basis he was able to calculate the different ways in which heat would flow depending (1) on the shape of a body, how much heat is applied to it, and where that heat is applied; (2) on the capacity of the body to absorb heat; and (3) on how quickly or slowly heat moves through the body depending on the temperature differences.

Fourier discovered his law by imagining an infinitesimally small box-shaped cell and supposing that the temperatures would differ infinitesimally across each of the three dimensions of the cell: front to back, bottom to top, right to left. The law he derived from this imaginative scenario was the following equation:

$$\partial T/\partial t = [K/C \cdot D](\partial^2 T/\partial x^2 + \partial^2 T/\partial y^2 + \partial^2 T/\partial z^2)$$

From here Fourier envisioned different ways in which an infinitesimal cell could communicate its heat flows to its tiny neighboring cells, how they would respond, and how they would in turn transmit heat to their neighbors, starting the process over again. In some situations, there could be loops where heat flows back into the original cell, producing an increasingly amplifying feedback cycle. Fourier used integral calculus to add up the interactions among the infinitesimal cells. He thereby determined how this would produce different kinds of macroscopic heat flows under a wide variety of circumstances. His innovative ways of using integral calculus to accomplish this task now bears his name: "Fourier analysis."

Fourier's law of heat flow had the form of a partial differential equation, so it is open to a remarkably diverse set of outcomes (or "solutions") under varying conditions. In effect, finding solutions amounted to adding up (integrating) each of the infinitesimal cells as one moved across a macroscopic body from the locations where heating sources were being applied to the locations where heat escaped from the body. He used his own Fourier analysis to solve the equation under many different kinds of conditions. A century and a half later, climate scientists would use high-speed computers to add up the contributions of the tiny components (cells) in the atmosphere and oceans to deal with conditions too complex to be solved by hand using Fourier's methods.

Among many other things, Fourier investigated how his law applied to heat passing through panes of glass. Experiments revealed, for example, that more heat was retained when air was contained between the panes of glass than when the intervening space was a vacuum. He observed: "In general the theorems concerning the heating of air in closed spaces extend to a great variety of problems. It would be useful to revert to them when we wish to foresee and regulate temperature with precision, as in the case of green-houses, drying-houses, sheep-folds, work-shops, or in many civil establishments such as hospitals, barracks, places of assembly."[3] Despite what this passage may suggest, providing a theoretical basis for utilitarian improvement of human industries was not Fourier's primary objective. His ultimate goal was to understand the dynamics of heat flow throughout the earth. This was "the key motivating factor in all of Fourier's theoretical and experimental work on heat."[4] Implicitly at least, Fourier's theorizing led in the direction of understanding a good that transcends and encompasses the satisfaction of human wants and needs, namely, the good of the planet earth itself. This constituted what Lonergan would call a shift from commonsense and descriptive (i.e., utilitarian) concerns to explanatory concerns.[5] This was only implicit in Fourier's work, but it would become explicit in the work of climate scientists and activists late in the twentieth century.

Using his law of thermal flow as his point of departure, Fourier identified three sources of heat that impinge upon the surface of the earth: the sun, the subterranean heat inside the earth, and the "temperature of space" due to the "innumerable stars, in the midst of which is placed the solar system."[6] To a first approximation, the earth will heat up until it reaches a steady state where it radiates back out as much heat as it receives from the three sources. In the second approximation, Fourier took into account the unequal heating by the sun due to latitudes and the earth's periodic rotation; these would introduce additional dynamical terms. To a third approximation, he took into account the earth's enclosure in an envelope of atmosphere.[7] This added a further complication because "heat in the state of light finds less resistance in penetrating air, than in repassing into the air when converted into non-luminous heat."[8] In other words, once incoming sunlight warms the earth, that sunlight is changed into a different kind of nonluminous heat that does not pass back out of the atmosphere as easily. Later scientists would discover that when sunlight heats earthly bodies, they radiate what we now call infrared radiation (nonluminous heat), more of which is absorbed than is transmitted back out through the atmosphere.

Taking still more factors into account added still further complications. Fourier's law worked for solid and liquid bodies, but not for gases—partly because gases expand and contract much more vigorously when heated or cooled. "As a result," Fourier observed, "it is difficult to know how far the atmosphere influences the mean temperature of the globe; and in this examination we are no longer guided by a regular mathematical theory."[9] In order to account for the effect of the atmosphere on the earth's heating and cooling, therefore, he relied upon the analogy of air trapped between panes of glass to make a general descriptive analysis.

Throughout this account, Fleming makes an especially concerted effort to show that Fourier was *not* the person who first predicted that the atmosphere would heat up when the quantities of carboniferous gases increase. He likewise emphasizes that Fourier did not regard the atmosphere or human activities to be major determining factors in the heat dynamics of the earth. Rather, Fourier held that the extremely cold "temperature of space" would play a much greater role because it would determine how fast heat would radiate away from the earth. Finally, although Fourier mentioned greenhouses, he did not explicitly compare the earth to a greenhouse, contrary to a frequently repeated opinion. Nevertheless, the implications were there waiting to be picked up by later scientists. In 1884 for example, William Ferrel relied on Fourier's work to "develop equations for the distribution of sunlight over the globe" and then showed that half of the heat absorbed by the "diathermanous" atmosphere would be radiated back to the earth. He calculated that the mean atmospheric temperature would be 15.4°C (59.72°F). This was in fact the observed mean temperature obtained at that time by the much more advanced worldwide meteorological reporting system. Even Ferrel, however, was not concerned with global warming. For him, the analysis of the atmospheric heating was necessary to explain why the average earthly temperature was not much lower than it in fact was, since he calculated that it would have been –96°C (–140.8°F) without the atmospheric factor.

Beyond the points emphasized in Fleming's narrative, there is a second feature of Fourier's theory of heat dynamics that had ramifications for the future science of climate change: his theory made clear just how complex climate dynamics are. His ultimate goal was to understand the complete dynamics of the heating of the earth, complete with an explanation of earthly climate dynamics, but he did not accomplish that goal in *The Analytic Theory of Heat* itself. Since his law of heat flow had the form of a partial differential equation, its actual physical manifestations would vary dramatically under countless different sets of conditions. Even though he applied his law to relatively simple and symmetric setups in *The Analytic Theory of Heat*, the solutions were already quite complex. As the conditions to which the law was applied became even more varied, the solutions grew ever more difficult. When Fourier tried to determine how heat would flow in atmospheric gases, the calculations became unmanageable. Fourier also recognized that applying his equation to all the other complex earthly conditions would be even more daunting. He explicitly listed the many complex factors that would have to be taken into account: the different amounts of heat radiation reaching different parts of the planet, how much radiation is reflected back from different types of surfaces (soil, rock, water, ice), the sources of heat from the earth's interior as well as from the sun, the source of cooling by outer space, the alternations in heating and cooling due to the earth's rotation, the differences between land and water heat capacities, the effects of the varying air densities at different heights of the atmosphere, the effects of air and water currents, and so on. This diverse and large number of interacting factors was beyond the considerable analytic prowess of even Fourier, and he did not fulfill his ambition to furnish a complete mathematical description of the heat system of the earth. Ultimately, it would require the invention of supercomputers and complicated processing software to provide a scientifically credible account of atmospheric dynamics and climate change.

Even though Fourier himself did not fully realize his dream, his equation has been used by countless atmospheric and climate scientists to the present. His work was indispensable in arriving at the sophisticated computer models that formed the basis for contemporary estimates of human-caused climate change, but Fourier himself did not envision the degree to which human industry and commerce could impact climate.

Chapter 12

The Growth of the Science of Atmospheric Warming

> Charles David Keeling . . . loved chemistry, and he loved the outdoors. . . . He was committed to . . . the laboratory, but he spent all the time he could traveling mountains and woodland rivers. . . . Keeling's work was one example of how geophysics research often rested on love of the true world itself.
>
> —Spencer R. Weart, *The Discovery of Global Warming*

This chapter sketches the complementary narratives of Fleming and Weart regarding the dialectical history spurred by questions about the growth and recession of ice ages. Once the existence of ice ages had been scientifically accepted, uniformitarianism crumbled. In the wake of its collapse, scientists sought alternative theories that retained its explanatory powers but also met the further questions posed by the facts about changing climates. It would be many decades before the explanatory power of Fourier's work was recognized. At this point in time, climate change was understood as the oscillation between ice ages and interglacial periods over long epochs of time. This was not yet climate change understood as anthropogenic global warming. In fact, at this time scientists were more concerned about whether the earth might be heading into a new ice age, what this might augur for human well-being, and whether anything could be done about it.

Fleming and Weart pick up their narratives with the attempts to provide astronomical explanations of the ice ages, which culminated in the Croll-Milankovitch hypothesis about minute changes in the earth's orbit contributing to changes in the earth's temperature. They then turn to the advances made by Tyndall's and Arrhenius' work on the effects of CO_2 emissions on the atmosphere, though their work was not conducted with anthropogenic global warming in mind. T. C. Chamberlain then explicated a CO_2 cycle that implicitly pointed to global warming, but it quickly fell out of favor due to the early state of scientific data available at the turn of the twentieth century. Guy Stewart Callendar was able to rescue the explanatory relation between CO_2 and changes in the temperature, eventually convincing the scientific community that human-caused CO_2 emissions could raise or lower the earth's overall temperature. He paved the way for Roger Revelle and Hans Seuss to argue that such emissions were outrunning the capacity of nature to reabsorb harmful radiation. This chapter concludes with the work of Charles David Keeling in the mid-twentieth century, who first identified what we now consider climate change, though its imminent danger was not readily apparent to scientists of his day.

Astronomical Explanations

Numerous scientists attempted to work out several different types of explanations for the cycles of ice ages. William Herschel discovered that some stars exhibit cycles of varying luminosity, and he suggested that the sun might also have a long cycle leading to a cyclical heating and cooling of the earth.[1] Weart observes that nuances were added to this proposal much later by Charles Greeley Abbot, who attempted to show that a corresponding sunspot cycle would lend support to the idea of variations in solar luminosity. Others suggested that periods of heightened volcanic eruptions would release a "volcanic fog" blocking sunlight and inaugurating cooling and glaciation. Still others proposed that the extremely cold water in the depths of oceans might somehow flow northward and begin to form ice accumulations.[2]

One hypothesis that gained several adherents posited variations in the earth's orbit around the sun. Isaac Newton already knew that the earth's orbit was not an exact, unchanging ellipse; instead, its orbit varied as the perihelion precessed minutely and slowly. In 1824, Jens Esmark presented a more complex analysis of how variations in the earth's orbit and the inclination of its axis toward the sun could have caused the kind of periodic warming and cooling needed for glacial cycles. In the 1860s, James Croll advanced upon the work of Esmark. He studied the works of astronomical physicists Joseph-Louis Lagrange and Urbain Le Verrier, who both greatly advanced the mathematics of Newtonian astrophysics. They had calculated how the varying locations of the moon and planets would reinforce or diminish the gravitational pull of the sun on the earth, thereby causing deviations from a perfectly elliptical orbit. Croll used their conclusions to argue that these long-term variations in the orbit of the earth would produce oscillating cycles of cooling and warming lasting over tens of thousands of years. He further argued that as colder winters produced more snow, this would reflect sunlight, further reducing temperatures so that this feedback mechanism would accelerate glaciation.

Le Verrier's calculations were not complete enough to carry Croll's work further. Numerous scientists, including Croll, realized that these variations would be too slight to produce the warming and cooling effects required for known glacial cycles. Others—especially Milutin Milankovitch—carried Croll's hypothesis forward into the twentieth century with greater precision. Between 1912 and 1941, Milankovitch carried out exacting calculations using advanced mathematics. He published several elaborate patterns of variations in orbital distances and showed how these would cause corresponding variations in the sunlight received by the earth over several cycles as short as 21,000 and as long as 650,000 years in length. His publications also showed correlations between the variations in sunlight and ice ages. However, the hypothesis that these variations could be causes of the ice ages was not accepted in his lifetime. The variations in sunlight that he calculated were regarded as too small to affect glaciation.[3] Many decades later, however, with some of the most advanced data on glaciation (for example, measurements from different layers of mud and ice cores extracted from the depths of oceans and glaciers) and advances in computer modeling, the Croll-Milankovitch hypothesis gained new respectability.[4] In particular, as computer modeling revealed complex dynamic systems that were extremely sensitive to minimal changes in initial conditions, the idea became plausible that the miniscule variations calculated by Croll and Milankovitch could have caused dramatic glacial changes. Nevertheless, during Milankovitch's own lifetime, his hypothesis joined all the other discarded attempts to explain the ice ages. As Weart put it, "Sooner or later, every prediction failed."[5] All failed, that is, until research into the radiation capture by water vapor and carbon dioxide were pursued in depth.

Discovering the Role of Carbon Dioxide

Fourier and later Ferrel speculated that atmospheric air absorbed what we now call infrared radiation at rates that were higher than absorption rates for other wavelengths of light, but it fell to John Tyndall to measure and establish this experimentally. An exceptionally talented scientist, Tyndall trained with Michael Faraday and made major contributions to many scientific fields. In 1859, he began his experiments on the radiative properties of various gases. He developed new, extremely sensitive instruments and procedures, and he demonstrated that "the absorption of thermal radiation by water vapor and carbon dioxide was of importance in explaining meteorological phenomena" including "variation of climates in the past" associated with ice ages.[6] In fact, he showed that water vapor molecules absorb 1,600 times more heat than other atmospheric gases such as oxygen, nitrogen, and hydrogen.[7] Water vapor was at the center of Tyndall's attention as the factor responsible for maintaining the average temperature of the earth, more so than carbon-based molecules.[8]

Svante Arrhenius took Tyndall's line of research several steps further. But Fleming goes to lengths to show that claims about Arrhenius as the "father of the greenhouse effect, even of global warming," are "misleading and incorrect."[9] Fleming argues, instead, that Arrhenius made important contributions to environmental science that would eventually support such theories, but he did not do so himself.

Like Fourier, Arrhenius harbored the ambition to develop a " 'cosmic physics'—the physics of the Earth, sea and atmosphere."[10] And like Fourier, his work moved science even further toward what Lonergan called "explanatory" science, that is, science as dedicated to understanding how vast ranges of data are related to one another.[11] Arrhenius therefore sought an explanation for the onset and retreat of ice ages within the ambit of his cosmic physics. Like his predecessors, however, he was quite concerned that the earth might be heading back into an ice age, not that we might be in the midst of an unprecedented epoch of global warming. This is why Fleming discounts him as the father of the greenhouse gas effect.

Arrhenius did little original experimental work. Rather, his contributions came with the theoretical application of "physical and chemical principles to existing observations."[12] Relying in part on Fourier's work, he constructed what for his time was a fairly complex mathematical model of global temperatures. In that model he gave more prominence to the role played by carbon dioxide (CO_2) than Tyndall did. Yet he still incorporated the radiation absorption properties of water vapor into his model and identified the possibility of complex feedback cycle interactions between water vapor, CO_2, and heating.[13] His model enabled him to calculate the temperatures over a wide range of latitudes and for different amounts of atmospheric CO_2. He did so with the expectation this would lead to an explanation of the cooling and warming responsible for the ice ages. In 1895 he presented a lecture "in which he suggested that a reduction or augmentation of about forty percent . . . might trigger feedback phenomena that could account for the glacial advances and retreats."[14] This was surprising, since CO_2 was known to be an extremely small component of the earth's atmosphere (0.04%). Since his concern was to explain the rise and fall of the ice ages, he was equally interested in the *reduction* as well as the increase of atmospheric CO_2.

He was quite aware that he had to incorporate many approximations into his mathematical model, given the state of science in 1895. He knew, for example, that the spectroscopic measurements for absorption of infrared radiation by CO_2 were accurate only for a limited range of its wavelengths. Even neglecting the numerous variables he was not able to incorporate into his model, Arrhenius still had to perform long series of "tedious hand calculations." Without high-powered computers, significantly greater complexities could not be taken into account.

In 1903, Arrhenius published a major textbook that elaborated his theoretical work on the physics of the earth, sea, and atmosphere, but "it was not widely read; it was a textbook for a discipline that did not [yet] exist."[15] In 1908, he published a more popular work, *Worlds in the Making*, which reached a wider audience. In this book he compared the atmosphere to a greenhouse, explaining how solar radiation can enter more easily than "dark heat rays" (infrared radiation) can escape. As Fleming notes, however: "The atmospheric greenhouse effect is even more complex than this, and a simple 'hot-house' theory of planetary atmospheres is not accepted today by atmospheric scientists."[16] Nevertheless, by the time of the publication of *Worlds in the Making*, Arrhenius "had become concerned by the rapid increase in anthropogenic carbon emissions."[17] Even though he recognized that this could mean a rapid increase in atmospheric temperature, still he did not predict a dangerous level of global warming. Relying to some extent upon the work of Arvid Högbom and T. C. Chamberlin, he thought instead that the world was likely headed toward a new ice age.[18]

Högbom, Arrhenius' friend and colleague, added yet another complicating variable into the newly developing science of the relations between atmosphere, land, and sea. He calculated the amount of carbon captured in limestone and other sedimentary deposits: "We find that about 25,000 times as much carbonic acid [i.e., CO_2] is fixed to lime in the sedimentary formations as exists in free air."[19] He proposed that in some geological epochs, CO_2 would be drawn out of the atmosphere into rock formations, thus cooling the earth and initiating glacial growth. In other periods, carbon would be released into the atmosphere, possibly through the decomposition of silicates and volcanism,[20] heating the atmosphere and shrinking glaciers. He did not think that the amount of CO_2 being generated by human uses was at all significant. Arrhenius relied upon Högbom's research, but he was not able to incorporate carbon capture into his model.

The person who most advanced the state of the science after Arrhenius was T. C. Chamberlin. Like Arrhenius, Chamberlin was interested in discovering relationships among the wide range of phenomena. He investigated the formation of the earth, sun, and solar system, the relationships between geological formation and the atmosphere, as well as the epochal variations in glaciers, temperatures, humidity, coal deposits, and fossilized species as evidenced in the geological record.[21]

Chamberlin built upon Arrhenius' research while he was working in 1896–1897 on his theory of glacial and climate cycles, adding his own chemical analyses along with accepted theories regarding continental formation. He argued that rock erosion would increase the exposure of calcium-containing rock to atmospheric CO_2. Chemical reactions would accelerate the capture of CO_2 in bicarbonate sedimentary deposits. This in turn would lead to a gradual lowering of global temperatures and the growth of glaciers. Erosion rates would be high during periods of continental uplift and mountain building, but as the continents were eroded and the bicarbonate rocks sedimented below sea level, the decrease in temperature would slow down. This would be followed by several processes that released CO_2 into the atmosphere: decay of organic material, discharge of CO_2 from cooler waters, increase of sulfates at the expense of carbonates in the oceans, and water vapor feedbacks. The combination of all these processes would lead to increased capture of infrared radiation. As the CO_2 and water content of the atmosphere increased, temperatures would begin to rise again. The glaciers would retreat, exposing more rock surface to erosion and thus beginning to reinitiate the cooling part of the cycle.

This is but a cursory summary of Chamberlin's remarkable analysis of this complex cycle.[22] As Fleming puts it, "Perhaps Chamberlin's greatest contribution was his understanding of the interconnections of the Earth's dynamic systems. Currently, one of our most difficult scientific challenges is to understand the complex stabilizing and destabilizing forces operative in the Earth system, many of which were first described by Chamberlin."[23] Chamberlin's account of the carbon cycle seemed destined to become a new privileged position, as Fleming puts it, because of his meticulous attention to detail and incorporation of so many dynamic systems. Within twenty years, however, his theory of the carbon cycle had fallen out of favor. Given the spectroscopic measurements available in 1905,[24] it was thought that the amount of CO_2 in the atmosphere already absorbed the maximum of infrared radiation that it could,[25] so any additional CO_2 could not absorb extra radiation and could not produce any further increase in temperature. As a result, instead of recognizing Chamberlin's theory as the privileged position, scientists instead pursued numerous competing hypotheses to explain the rise and fall of ice ages. Among these, William Jackson Humphreys argued in 1913 that the major cause of glaciation was periodic volcanic eruptions whose dust blocked sunlight and led to global cooling.[26] In that same year Chamberlin himself came to regret his overenthusiastic acceptance of the theories of Arrhenius and Högbom.[27]

Fleming remarks, "In the first half of the twentieth century, most scientists did not believe that increased CO_2 levels would result in global warming."[28] Beginning in 1938, however, Guy Stewart Callendar began to reconsider the case on two fronts. Over the span of two decades he published a remarkable series of research papers that gradually elaborated a more precise model connecting CO_2 and atmospheric temperature. He carefully calculated the rise in CO_2 in the atmosphere between 1860 and 1938 and "pointed out that fuel consumption had generated [150 billion] tons of carbon dioxide" over this period of time.[29] He then developed a model of the relationships between "downward sky radiation" (that is, infrared heating) and CO_2 concentrations for the different atmospheric pressures at varying altitudes. Most importantly, over several years he assembled a number of spectroscopic measurements independently done by others to show that there was, after all, significantly more absorption by CO_2 of a band of infrared radiation that had not been previously known, with a wavelength of eight to twelve microns. He was able to show that CO_2 absorption of infrared radiation over all its wavelengths would eclipse that of water vapor. This overturned the long-held assumption to the contrary.

Some of the infrared radiation absorbed by CO_2 would also be reemitted back toward the earth, warming the land as well as the atmosphere. On the basis of his model, Callendar argued that the increase in CO_2 accounted for 60 percent of the then-observed half-degree Celsius increase in global temperature since 1900, and that a doubling of the amount of CO_2 would lead to a further increase of 2°C. Fleming remarks that a "discussion of [Callendar's 1941] article at the Royal Meteorological Society revealed significant changes in opinion" regarding the role of CO_2 in atmospheric heating.[30] By 1961, three years before the end of his life, Callendar had completed an impressive body of research and concluded that "the trend toward higher temperatures was significant . . . [and] that increased use of fossil fuels had caused" this rise.[31] Nevertheless, Callendar was still concerned with the onset of a new ice age and thought "warmer was still better."[32]

Callendar's research was the tipping point that human-caused CO_2 emissions was a live option as the major source of global warming. While numerous questions remained, it became a privileged position that scientists now considered as possibly worth further exploration. "Scientists, inspired by Callendar, began to investigate in greater detail the linkages between rising CO_2 levels and rising temperature."[33]

The Complex System of Global Heat Dynamics

One of the most important scientists to provide greater detail to the relationship between CO_2 emissions and global warming was Gilbert Plass. In his 1956 article, Plass significantly refined Callendar's figures by calculating the variation of CO_2 absorption of infrared radiation for each kilometer of altitude between 1 and 75 km.[34] He relied on more accurate measurements done after Callendar's work, including some that he himself conducted. He was able to show that while the amounts of

CO_2 at lower latitudes had almost completely saturated infrared absorption already, the same was not true at higher altitudes. The release of additional CO_2 would find its way into the upper levels of the atmosphere and increasingly impede infrared radiation into outer space. The result would be the heating of the atmosphere and the land (and, contrariwise, reductions of CO_2 would lead to temperature decreases). He acknowledged that parallel accurate measurements for H_2O were not yet available but used reasonable estimates to calculate the additional radiation capture contributed by water vapor. He also incorporated detailed calculations of the uptake of oceanic CO_2 by dissolved calcium carbonates into his model, as well as calculations for the different amounts of cloud cover, from pure sky to various degrees of cloudiness at different altitudes. Among the several major conclusions in his article, two stand out: (1) "If the total amount of CO_2 in the atmosphere-ocean system is reduced by a small amount from its present value . . . the climate must fluctuate between periods of large ice sheets and warmer periods . . . [so] there is no possible stable state of climate"; and (2) he endorsed Callendar's prediction that present rates of fossil fuel combustion are "increasing the average temperature by 1.1°C per century." He added that the warming trend would continue for several centuries, and that if all available fossil fuels were consumed in the next thousand years, "the corresponding temperature rise is 7.0°C."[35] Scientific consensus that the rate of warming was much faster was still decades away.

In order to incorporate so many variables, Plass used the new MIDAC high-speed digital computer to run his more complex and "more realistic models of radiation transfer."[36] He used his sophisticated computer model "to warn that the accumulation of carbon dioxide in the atmosphere from anthropogenic sources could become a serious problem in the near future."[37] Nevertheless, because Plass had to rely on several estimates that have since been revised on the basis of more recent research, at least one scientist (Lewis Kaplan) cast significant doubt on his findings in 1956.[38] Plass himself explicitly recognized the limits of his findings and that more research was needed: "*If* at the end of this century, measurements show that the carbon dioxide content to the atmosphere has risen appreciably and at the same time the temperature has continued to rise . . . [*then*] it will be firmly established that carbon dioxide is *an* important factor in causing climatic change."[39] The most advanced scientist in this field still did not think in 1956 that human-caused global warming was a serious concern.

Weart remarks that even after Plass' work at least one important further question remained, one that Plass himself recognized but lacked sufficient information to incorporate into his model: "If Plass had shown that the facts of infrared absorption did not rule out greenhouse warming, another weighty objection to the theory remained. Wouldn't the oceans simply swallow up whatever extra CO_2 we humans might put into the atmosphere?"[40]

Beginning in 1955, Roger Revelle collaborated with Hans Seuss on the relationships among atmosphere, oceans, and CO_2. Revelle studied surface-to-depth currents, while Seuss used radiocarbon dating to calculate the ratio of CO_2 in the atmosphere that actually did derive from the burning of ancient fossil fuel sources. Together they endeavored to determine "the carbon cycle," that is, how much CO_2 produced by human and other sources is captured by the atmosphere, the oceans, living organisms (e.g., plants), and the lithosphere. They estimated that oceans captured twice as much CO_2 as the atmosphere but that "carbonates in sediments were two to three orders of magnitude larger" in their capture rates than the oceans.[41] Their initial work led them to conclude that the oceans could absorb all of the CO_2 being produced by twentieth-century human civilization, but "one day [Revelle] realized that the peculiar chemistry of seawater" would function as a buffering mechanism and "would prevent it from retaining all the carbon that it might take up."[42] Even though most of the CO_2 would be absorbed by the oceans, it would soon be released because of the buffering chemistry.

Each scientist brought his own specialty to this collaboration. Seuss had developed a technique of radiocarbon dating to determine the rate at which fossil fuel carbon was being added to the atmosphere. In 1957, they published what has become a most famous research article. The discovery of the buffering effect was inserted late into the article and did not initially stand out as a major finding. Yet the line that has made their article famous reads: "Human beings are now carrying out a large scale geophysical experiment of a kind that could not have happened in the past nor be reproduced in the future. Within a few centuries we are returning to the atmosphere and oceans the concentrated organic carbon stored in sedimentary rocks over hundreds of millions of years."[43] Neither Revelle nor Seuss was particularly alarmed when the two wrote these sentences; they thought that the effects would not appear for "a few centuries." Their work only established that the oceans would not be able to absorb all the CO_2 produced by humans. Neither they nor anyone else knew for certain what amounts of CO_2 were being released. The numbers that Revelle and Seuss used in their article were chosen from wildly varying estimates. As Fleming put it, "Clearly carbon fluxes were not well known"[44]—not, that is, until the work of Charles David Keeling.

While a postdoctoral student in the 1950s, Keeling became interested in the question of the CO_2 content of the atmosphere. Revelle heard of his work and secured a research position for him on a grant he had obtained from the US government in cooperation with the International Geophysical Year (IGY, 1957–1958). Revelle had an agenda, but Keeling had his own ideas: he wanted to see if it were possible to establish a reliable baseline measurement of the amount of CO_2 in the atmosphere. The difficulties of screening out random sources of CO_2 were considerable. Variations in winds could carry wafts of CO_2 from forests or factories or vehicles; even human breaths could disturb the accuracy of measurements.

To make his CO_2 measurements, Keeling developed new, highly sensitive instruments. They were so sensitive, however, that he needed to isolate them from randomly fluctuating sources. He came up with the idea of setting up his instruments at the observatory isolated atop Mauna Loa in Hawaii, and in Antarctica.[45] In 1958, he began the continuous recording of meticulous measurements of CO_2 levels at the Mauna Loa Observatory, producing what is now famously known as "the Keeling curve." The curve oscillates over the course of each year, but the peaks for each year rise continuously upward.[46] Keeling called his research "the first unmistakable evidence of atmospheric CO_2 increase."[47] While the work of Callendar had established a privileged position for the link between CO_2, temperature fluctuations, and glaciation, Keeling's research sealed another part of the deal: that the amounts of CO_2 in the atmosphere were in fact steadily increasing, and that human activities were a likely source. Hence the work by all the others took on much more profound significance. In 1963, Keeling and others participated in a conference and issued a report "*suggesting* that the doubling of CO_2 projected for the next century could raise the world's temperature by more than 4°C (more than 6°F). They warned that this *could* be harmful: for example, it *could* cause glaciers to melt and raise the sea level enough to submerge important coastal areas."[48] The report was one of the first to use the language describing what is now known as climate change—not only global warming but also associated consequences that would be devastating for natural and human ecosystems.

Yet the report only used language of *possibilities*: "suggest," "could," and so on. In fact, three years later a special committee of the National Academy of Sciences pronounced that "*there was no cause for dire concern*" although "human activity *could* influence climate" and should be "watched closely."[49] In order to make a solid case for the reality of the link between increasing CO_2, the rise in global temperatures, and the consequences for the changing climate, further work of climate modeling by advanced computers would be required. We turn to this in the next chapter.

Conclusion

This chapter has followed scientific developments over the course of a century as documented in the work of Fleming and Weart. Their narratives reveal the complex route that science took over the course of that century. Even though neither historian knew about Lonergan's work, what he called the self-correcting process of knowing is prominent in their narratives. This period began with the data about ice ages, and the question about their causes. It moved through a series of insights, hypotheses, mathematical models, and further questions in search of an explanation based on the earth's orbit around the sun. These efforts seemed to reach a dead end but stimulated further questions about radiation capture by atmospheric gases. The insights and experiments that responded to these questions led to still further insights, questions, and experiments by Arrhenius, Högbom, Chamberlin, Callendar, Plass, Revelle, Seuss, Keeling, and numerous others. Notably, this was not a straightforward linear path of progress. Several hypotheses that were promising eventually faced questions that could not be answered within the limits of theoretical and experimental knowledge available at the time. Some of those hypotheses would be revived and revised later in the history of climate change science.

Over the course of the century, scientists gradually began to understand that the intelligibility of climate and global heating is incredibly complex. They gradually discovered some of the connections among phenomena involved in this complexity—including some connections with human activities—but even by 1960, they realized there were a great many questions still requiring answers.

As with other realms of ethical concern, climate change ethics must begin with accurate, factual knowledge of the situation. The further questions and answers about the right things to do depended upon that knowledge. By the end of the period covered in this chapter, scientists still did not have the degree of certainty about the facts of global warming and climate change they needed to make objective ethical judgments and propose ethical courses of action regarding climate change. That situation would change dramatically in the next few decades. This is the subject of the next five chapters.

Chapter 13

The Rise of Computer Modeling

Computer Modeling as the New "Privileged Position"

We have seen how scientists in the late nineteenth and early twentieth century, motivated by the need to account for the rise and fall of ice ages, began to develop tools for explaining the correlation between the earth's atmosphere and its temperature originally explored by Fourier. Eventually, human-caused CO_2 emissions were identified as a significant contributing factor to changes in global temperature, but research up to the early and mid-twentieth century had yet to recognize (1) that the overall change was in the direction of global warming rather than global cooling; (2) that the threat was present and in the immediate future; and (3) that this global warming would have numerous and varied future consequences for nature and humanity. We have followed Fleming's account of the shifting privileged positions that guided scientific research. We also saw that the recognized immensity of variables prevented the development of theories, explanations, and equations adequate to the complex phenomena we now call climate change. To achieve such theories, scientists needed the aid of advanced supercomputers; hence, the privileged position of computer modeling emerged.

According to Fleming, Gilbert Plass inaugurated a new privileged position in climate science in which computer modeling and complex algorithms took center stage.[1] Subsequent computer models used increasingly more accurate numbers as inputs to replace Plass' estimates. They were also able to incorporate more factors, such as the changes in reflection of sunlight by snow and ice (the "albedo" effect). In support of his claim for the emergence of this new paradigm in climate science, Fleming gives only a terse survey of the work between 1956 and 1974. Weart, however, goes into much greater detail about the history of the advances in computer modeling of climate change, so this chapter relies largely on his narrative.[2]

In 1922, Lewis Fry Richardson had the insight to model the earth by dividing it into a grid of cells, much like Fourier had done when developing his law of heat flow a century earlier. For Richardson, however, the cells were not infinitesimal but small and finite. Each was assigned its own set of numbers corresponding to pressure, temperature, and so on. Richardson "would then apply the basic physics equations [such as Fourier's and] calculate, for example, wind speed and direction according to the difference in pressure between two adjacent cells."[3] The numbers in each cell would be modified by its neighbors and modify them in turn in complex feedback patterns. Richardson recognized, however, that the number of calculations required to carry this out for even a relatively short period of time would be practically impossible: "He doubted they would be able to compute weather faster than it actually happens."[4]

This was decades before the invention of high-speed computers, the development of which was accelerated by attempts during World War II to decipher Nazi encoded messages. Following the war, John von Neumann worked on computer simulations of nuclear explosions. He then began to advocate applications of computers for weather prediction, which were supported financially by the Navy in hopes of gaining military advantages. He recruited Jule Charney to work on the project. "In 1950 Charney's team solved Richardson's problem: for a chosen day they computed something roughly resembling what the weather had actually done."[5] It would be more accurate to say that his team "sort of" solved Richardson's problem, since they solved it under a great many restrictions. To give just a few examples, they could only compute with accuracy what would happen in the next twenty-four hours, and only for restricted regions, and they had to use averages to smooth out mountains and valleys into an approximately smooth-surfaced globe. Nevertheless, the model was regarded as a great success even under these limitations. Their efforts in turn generated demands for an enormous increase in collecting atmospheric data, which were then fed back into computer models. The computers were slow by contemporary standards, and the predictions were produced at just about the same time that the weather actually occurred. By 1955, however, they made significant advances in speed of calculation, which would continue to improve over the next decades.

Their goal was to develop a General Circulation Model (GCM) for the entire planet. In 1955, Norman Phillips designed and ran what was called "the first true" GCM. It modeled twenty days of weather fairly accurately, displaying such phenomena as wind eddies and jet streams comparable to those actually observed. But after twenty days it started to model bizarre patterns that deviated significantly from actual weather. Phillips and many others worked to build improved models with more and more accurate data. They had some successes, but the odd deviations kept coming back.

In 1961, Edward Lorenz was working with his computer model at MIT. "One day he decided to repeat a computation in order to run it longer from a particular point. His computer worked things out to six decimal places, but to get a compact printout he had truncated the numbers, printing out only the first three digits."[6] At first the repeat version resembled the initial run, but, to his and everyone else's astonishment, the truncated model soon deviated wildly from the original. "A difference in the fourth decimal place [i.e., only 1/10,000] was amplified in the thousands of arithmetic operations, spreading throughout the computation to bring a totally new outcome."[7] Lorenz went on to investigate in depth why his model and others like it would behave in such bizarre ways. He eventually proclaimed that "precise very-long range forecasting would seem to be non-existent."[8] Years later Lorenz' discovery was made famous to the general public in James Gleick's popular book *Chaos*.[9]

The association between Lorenz and the word *chaos* was unfortunate—later generations continuing his area of studies now prefer to refer to their field as "complex dynamic systems theory" (or just "dynamic systems theory"). Among the things discovered by Lorenz and explored by others is how highly sensitive the stabilities of certain systems are to minor deviations because of the phenomena of feedback and self-amplification.[10] Some systems have only two stable states, and miniscule perturbations send them rapidly toppling from one stable state into another. As computer simulations of climate advanced, many of the models possessed just such dynamisms: a climate might exhibit only a few stable states, and relatively small influences would cause it rapidly to shift to a different state. As Weart puts it, "Most scientists agreed that climate has features of a chaotic system, but they did not think it was wholly random . . . tornado seasons [for example] came on schedule. That type of consistency showed up in computer simulations constructed in the 1980s. Start a variety of GCM runs with different initial conditions, and they would show random variations in the weather patterns. . . . But the runs would converge when it came to the global temperature."[11] This means that dynamic systems are not wholly chaotic after all. Lonergan would add that predictability is the intelligibility characteristic of what he called "systematic processes," but that this is not the only kind of intelligibility. Complex dynamic systems fall into the broad category of what he called "non-systematic processes." While nonsystematic processes lack predictability, they are not chaotic. They exhibit another kind of the intelligibility, the intelligibility of probability "from which actual frequencies may diverge but only non-systematically."[12] *This* is the "type of consistency" that characterizes the computer simulations and the real climate they model. Lonergan's identification of this alternative form of intelligibility will be shown to have great significance for environmental ethics in chapters 15 and 16.

The idea that the climate might have nonsystematically alternating stable states was not entirely new. Prior to the advances in computer modeling of the climate, Cesare Emiliani conducted ingenious measurements on suboceanic mud cores dating back three hundred thousand years. His remarkable measurements of CO_2 contents (and therefore temperatures) at different layers led him to reject a four-ice-age hypothesis in favor of many more numerous and irregular patterns of glaciation and melting. Remarkably, his findings corresponded closely with the variations in solar illumination calculated by Milankovitch. Although his findings were justifiably criticized for oversights, subsequent research by Wallace Broecker and others substantiated both his findings and their correspondence with those of Milankovitch.[13] Earlier, Broecker had speculated in his doctoral dissertation that within the data on glaciation "two stable states exist, the glacial state and the interglacial state, and that the system changes quite rapidly from one to the other."[14] This type of nonsystematic process had also been proposed much earlier in 1925 by respected climate expert C. E. P. Brooks, although on the basis of far less evidence.

The topic of multi-stable states of climate was discussed at the 1965 conference on "Causes of Climate Change." Roger Revelle stated the consensus at the end of the conference: "Minor and transitory changes in the past may have sufficed to 'flip' the atmospheric circulation from one state to another." Among the candidates for such minor changes were Milankovitch's slight variations in solar illumination. Even though these variations were slight, it was now accepted that the dynamic system of feedbacks in the earth's climate could have amplified them, thereby flipping from one state to another in a relatively short period of time (at this point they were thinking in terms of each period lasting a thousand years).[15] Since his calculations corresponded so closely with the data from mud and ice cores, the once discarded hypothesis gained new life.

The variations in solar illumination alone were not the direct cause of the ice ages. It was, rather, how the variations in solar illumination were amplified by the complex dynamism of the earth's climate, which caused the great variations in climate. Over the next decades, the web of dynamic relationships expanded ever further to include additional elements: the salinity of the sea, smoke, haze, aerosols, dust (from volcanic eruptions, for example), water vapor, clouds, and the reflection of sunlight by snow and ice. Each of these elements also has inherent, nonsystematic, and dynamic feedback dimensions of its own, further intensifying the complexity of climate dynamics.

As modelers attempted to incorporate these and other elements, new questions arose from various parts of the scientific community. For example, there was vigorous debate over whether the effects of haze and aerosols would tend to cool or warm the atmosphere and land. There were good scientific cases on both sides. Because they scatter and reflect incoming sunlight, there was a good case that they must be cooling agents. The specter of global cooling had continued to haunt climate science from the early theories of the formation of the planet, which claimed that it had formed as a fiery ball and had been cooling ever since. The original investigators of ice ages were convinced that the earth was heading into another ice age. Even into the latter half of the twentieth century a great many scientists expected that if there were to be any rapid "flipping" of climate, it would be from the present interglacial period into a new ice age. Even in that case, they did not think there was concern about cooling, since "rapid" change in climate states were thought to occur on the order of a thousand years. In fact, actual data showed that the average global temperature had been gradually cooling since 1940. Was the global climate really cooling?

James Hansen and his team of researchers took up this question. It was not easy to answer. The data on global temperature over the previous century was uneven in quality and had to be purified. In addition, potential causes of particular episodes of cooling had not been investigated thoroughly. Among other things, Hanson's team developed a computer model of a 1963 volcanic eruption and were able to show that the sulfates that were ejected brought about temporary cooling.[16] This led them to further investigate the role of other aerosols, especially chlorofluorocarbons (CFCs). The presence of CFCs in the atmosphere had begun to increase in the late 1940s, when they were introduced as propellants in spray cans and coolants in air conditioning units.[17] Hansen and his team showed that temporary increases in aerosols, including CFCs, caused a temporary cooling that masked what would have been the ongoing increase in rising temperature due to CO_2. This claim was corroborated by other scientists as well.[18]

Hansen's investigations of aerosols led to other research. In 1975, Veerabhadran Ramanathan demonstrated both that aerosols were powerful greenhouse gases and that they had deleterious effects on the ozone layer. In addition, methane (CH_4) and nitrates (especially N_2O)—both of which are emitted in large quantities in agriculture—were found to be even more absorbent of infrared radiation than CO_2.

Although Revelle had earlier shown that the buffering of oceanic chemistry would minimize the long-term retention of CO_2 in ocean waters, Hansen's team discovered another nuance. They modeled the increased concentration of CO_2 in the atmosphere and showed that the amounts taken into the oceans would likewise initially increase temporarily (even if they would eventually be released back into the atmosphere later). Therefore, actual increases in CO_2 released into the atmosphere were being masked by increased ocean uptakes. The releases back into the atmosphere would occur after a twenty-year time lag, along with their inevitable warming effects.[19]

Of special importance was the collaboration between Suki Manabe and Kirk Bryan that began in the late 1960s. Manabe established his reputation when he and his collaborators originally worked on modeling the complex convection currents that carried heat and moisture from the earth's surface to the upper atmosphere. They wanted to determine just how sensitive the climate was to minor inputs. In 1967, they used their model to see what would happen if CO_2 levels in the atmosphere were doubled. Their answer: global temperature would rise by 2°C. Weart reports: "This was the first time a greenhouse-effect warming calculation included enough of the essential factors to seem reasonable to many experts."[20] At the same time, Bryan had been developing a model of ocean currents and energy transfers. Subsequently they worked together to make their models interact. The initial versions had many limitations, but their combined efforts gradually began to provide a more realistic and much more complex GCM. In 1975, their "supercomputer ran for fifty straight days, simulating movements of air and sea over nearly three centuries." While it still lacked exact correspondence with the real climate, their findings were corroborated by the model of another team led by Warren Washington. Although Washington's algorithms were quite different from those of Manabe and Bryan, their results converged.[21] Subsequent teams improved upon these models, ever incorporating additional elements, new data, and new interactions. Weart assesses their outcomes as follows: "The results, for all their limitations, said something about the predictions of the earlier atmosphere-only GCMs. It turned out . . . that the oceans would delay the appearance of global warming for a few decades by soaking up the heat . . . [yet the more advanced] GCMs did not turn up anything to alter the predictions in hand for future warming."[22] These advanced GCMs revealed that rapid "flips" to different climate states (dramatically warmer as well as cooler) were not only possible but of shorter duration than was previously thought—perhaps mere centuries rather than millennia.

In the early 1970s, climate scientists also began to think about how biological systems might be intersecting with atmospheric and oceanic systems. They "noticed that there were additional components, even more complicated and scarcely studied at all" and in particular asked whether "living ecosystems were somehow essential parts of the climate system."[23] They began facing questions about how atmospheric carbon was being taken up by forests and fields, about whether increased CO_2 would stimulate plant growth into a feedback loop, and about whether that would release more water vapor into the atmosphere, and whether changes in land vegetation (e.g., through overgrazing) might be increasing the reflection of sunlight (albedo effect) and thus cooling the surface and affecting wind currents.

In particular, they began to face the question of whether "human activity could change vegetation enough to affect the climate." Up to this time, studies and concern about human-initiated deforestation had focused on "the sake of wildlife rather than climate."[24] Climate scientists now recognized that the amount of CO_2 being absorbed into the atmosphere according to Keeling's curve was only about half the amount of that gas being generated by human activities. How much was being converted into biomass and how were human activities affecting that? When computer modelers began to consider these questions, they realized how little data they actually had to work with. Their earliest crude models had to use assumptions with little empirical support. Scientists debated back and forth about the exact role of ecosystems in climate systems and about human impact on both.[25]

In effect, this became the question of how to integrate the advances in environmental science (discussed in part 1) into a larger framework. Gradually some data on CO_2 conversion to biomass were gathered and computer modeling began to take these into account. This is still an ongoing emerging dimension of climate science, however, and it points to the need for a framework capable of integrating these numerous developing sciences and their ongoing understandings of numerous dynamic systems. It is the kind of framework that Pope Francis has called an "integral ecology" in *Laudato si'* or that I called "the emerging good."

Impact on the Practice of Science

The watershed year for climate science was 1977. This was the year when, for the first time, a panel of experts from the National Academy of Sciences (NAS) "reported that temperatures *might* rise to nearly catastrophic levels during the next century or two" and that this was connected with human production of energy.[26] Weart observes that "in the choice between warming and cooling, scientific opinion was coming down on one side. . . . More and more scientists were coming to feel that greenhouse warming was the main thing to worry about."[27] Although much important research has continued up to the present day, 1977 was the year when the scientific community reached a consensus that global temperature was increasing, not cooling, and that this was serious.

The NAS declaration rested on the computer simulations of Manabe, Bryan, Hansen, Broecker, Washington, and others, along with the increasingly sophisticated data analyses of ice and mud cores that had been accumulated over the decade by many others around the world.

Over the course of slightly less than a century, the rise of climate science had an impact similar to that of the development of the science of ecology. Scientists had begun "to question the comfortable belief that climate was regulated in a stable material balance, immune to human intervention."[28] Complex dynamic systems were beginning to take the place of that long-held uniformitarian concept. Just as with ecosystems, "climate was the outcome of a staggeringly intricate complex of interactions and feedbacks among many global forces."[29] Weart writes, "Orbital changes, wind patterns, melting ice sheets, ocean circulation—everything seemed to be interacting with everything else. During the 1960s not only climate scientist researchers but also scientists in other fields and alert members of the public, were coming to recognize that the planet's environment is a hugely complicated structure. Almost any feature of the air, water, soil, or biology might be sensitive to changes in any other feature."[30] This change in the heuristic anticipation of what the world is like also impacted the very practice of science itself. While the traditional portrayal of scientific advance as the product of individual geniuses acting as "lone rangers" still persisted in popular imagination, that was no longer the reality for scientific growth. Well before 1977, science in general had started to become more of a collaborative enterprise within specialized fields. The growth of ecological and climate science pushed these boundaries further, and collaboration had to become interdisciplinary. This was highlighted in the NAS announcement, as with so many publications during the 1970s.

The reality of complex dynamic systems as well as the complex interdisciplinary investigations also forced a new understanding of scientific verification. No longer could scientific knowledge as such be regarded as the result of some unique "crucial experiment." "Scientists were giving up the traditional approach in which each expert championed a favorite hypothesis about one particular cause of climate change."[31] According to a 1969 remark by one authority, climate changes "are to be attributed to a complex of causes," and therefore "scientists needed one another."[32] Again, Weart: "If scientists were increasingly willing to consider that global climate might change greatly in the space of a century or so, it was because different pieces of the puzzle seemed to fit together better and better."[33] Weart uses the metaphor of a puzzle of many pieces in an effort to explain this, but I think it is possible to go beyond the metaphor and draw upon Lonergan's account of cognitional structure to clarify what was happening in the effort to piece together the many elements contributing to climate change. To this I will return in chapter 16.

From Scientific Complexity to Ethical Responsibility

Though a watershed, 1977 was hardly the end of scientific research on climate change. After all, the NAS declaration only stated that temperatures "*might* rise" in the next "century or two." While many scientists were individually convinced about stronger conclusions, most recognized that further research and modeling were needed before they could be persuaded that there were grounds to endorse stronger claims. They were aware that many further questions called out for answers. Were there other steadily increasing greenhouse gases (GHG) besides CO_2, methane, nitrous oxide, and aerosols? Was the combined rise in GHGs really the cause of the documented steady rise in global temperatures? Was the rise in GHG emissions truly due to human activity? Was there a tipping point that would set off a rapid and perhaps irreversible change in the state of the climate? Would the consequent changes in climate for humans and other living beings really be devastating?

Such questions stimulated a dramatic increase in the publication of peer-reviewed scientific articles on a wide range of topics related to climate change, and scientists "wanted more strongly than ever a centrally coordinated and generously funded program" of research.[34] It would take many more years of advances in data gathering, interdisciplinary synthesizing, and computer modeling to provide the needed foundations to form a scientific consensus around the answers to these further questions.

For almost a century, climate science research had been driven primarily by the desire for scientific understanding, although ethical considerations had always played some role. Climate scientists, after all, had to make responsible decisions about which lines of research were scientifically worth pursuing, and since their research for almost a century seemed to point toward the onset of a new ice age, ethical concerns about the potential impact of an ice age on natural ecosystems and human societies were at least at the margins of their thinking. By 1977, though, the results of the scientific research itself began to stimulate more serious ethical questions and concerns. These ethical concerns in turn motivated scientists to seek better understandings of global temperatures and their relationships to climate dynamics, in order to form better ethical precepts with greater confidence. As confidence in their ethical judgments began to grow, scientists realized that they needed to address the general public and the channels of political decision making. The consequences they began to consider could not be adequately addressed solely by individual decisions to use less energy or water. More and more, scientists gradually recognized that coordinated actions by institutions, businesses, governments, and nations would be required in addition to individual decisions. While the motivating force behind the previous century of climate research was the scientific desire to understand nature, 1977 marks the point where ethical and political concerns began to emerge as much more important motivations. From this point forward, the development of scientific, ethical, and political thought became intertwined in complex ways.

Weart narrates the many investigations and debates that engaged scientists trying to answer these questions to the satisfaction of the wider scientific community. Two years after the 1977 NAS declaration, scientists convened at the World Climate Conference in Geneva to discuss more recent findings. There they would agree that global warming might become a problem within their own lifetimes, not a century or two away. They also went on record saying that governmental policies *could* affect the rate of global warming.[35]

Nevertheless, what we now take for granted as the scientific consensus about climate change and the ethical responses proper to it did not begin to emerge until a decade later. In 1983, NAS released its report, *Changing Climate*.[36] In 1986 the British Climatic Research Unit "produced the first entirely solid and comprehensive global analysis of average surface temperatures." The analysis showed the same upward trend in surface temperatures that Keeling had measured for the rise in atmospheric CO_2.[37] In 1987, Broecker would write, "We must view [the greenhouse gas effect] as a threat to human beings and wildlife."[38] It was not until 1988, however, at the World Conference on the Changing Atmosphere that "a group of prestigious scientists" would pronounce that the world's governments *should* "set strict, specific targets for reducing greenhouse gas emissions."[39] And only in 1995 did the Intergovernmental Panel on Climate Change (IPCC) take the position that "the balance of evidence suggests that there is a discernible human influence on global climate." Finally, in 2001 the IPCC "bluntly concluded that the world was rapidly getting warmer and declared it likely that this was due mainly to greenhouse gases." The IPCC also explicitly connected this to drastic rises in sea levels and affiliated consequences.[40] By the time of the 2007 IPCC report, "scientists did feel more certain about some things," especially that "the effects of global warming were now upon us," not in the distant future. This report also connected this explicitly with extended droughts, severe storms, rise of diseases and pests, and threats to fisheries.[41]

It was during this period that scientists began to confront the ethical challenges posed by the results of their scientific research. Their knowledge of the facts about climate and its highly probable impacts were startling. Their predictions of the highly probable loss of environments, of arable land, and biodiversity, desertification, rising sea levels and increasing catastrophic storms, climate refugee migrations, to name a few, elicited strong feelings of dread about such outcomes. Both the facts and the feelings they elicited set in motion processes of ethical reflection seeking ideas about what could be done, what should be done, and the values that should guide actions. Those ethical questions and challenges began also to motivate further lines of research. These feelings and reflections made scientists aware of their ethical responsibility to increase public awareness and to engage in political processes. In doing so, they were pushed to respond to still further questions, questions that were outside the frameworks of their professional training. The need for some kind of a more comprehensive heuristic framework of the emerging good became palpable. From then on, the history of climate change became not just a scientific but a public affair. This is the subject of the next chapter.

Chapter 14

Going Public

Ethical Response to Climate Science

In *The Ethics of Discernment* I argued that the first step in ethical discernment and action (the first step in the structure of ethical intentionality) is correctly understanding the situation.[1] By 1977 climate science had revealed a situation of complex interactions among numerous dynamic systems. Any genuinely ethical response would have to rise to that level of complexity. Although it would be another decade before the majority of scientists reached a consensus about the dynamics of climate change and about the likely alternative future scenarios, by 1977 it was already clear that the production of GHGs by virtually every human endeavor on the planet was the major contributing factor to the complex pattern of climate changes. This meant that the ethical response would also have to be a complex coordinated effort involving virtually every human being. Today some climate change activists speak as though the ethical response is as simple as turning off a single circuit breaker. It is not.

The first step in ethical response to climate change had to be dissemination of the scientific findings. This chapter traces the history and challenges in that first stage of expanding from scientific investigations to ethical reflections and activities, namely: raising public awareness. Subsequent chapters will examine how ethical responses developed after a certain level of public awareness had been achieved.

From Scientific Facts to Ethical Discernment

The following is a summary of some of the major facts about our current climate situation that scientists hold with a high degree of probability.[2] These facts were derived from the scientific investigations and computer modeling discussed in the previous chapters:

- The earth really is warming.
- Increases in GHGs are the principal cause of this warming.
- The main source of increased GHGs is human activities of energy production and consumption (in industry, agriculture, commerce and transportation, domestic heating and cooling, entertainment, etc.).
- This remains true even after factoring in the many ways in which GHGs are taken out of the atmosphere by oceans, rock formations, and plant life.
- The capacity of some of these sources to absorb GHGs is being diminished by the increases of heat and by GHGs themselves, as well as human activities such as deforestation and depletion of ground watersheds.
- The earth's average temperature has already risen by more than 1.1°C since the Industrial Revolution, by more than 0.7°C since 1900.[3]
- If there is no change in the rate of GHG emissions, current models predict that the earth's average temperature will rise between 1.4°C and 2.7°C by 2100, depending upon the model and the assumptions about future rates of GHG emissions.[4]
- Such rises will increase desertification, diminishing available agricultural land.
- Increases in average temperature have already and, if unchanged, will continue to melt glaciers, ice sheets, and polar ice caps, reducing reflected sunlight and further accelerating warming.

- If unchanged, such melting will also result in a rise in sea levels of between eight to twenty inches, and changes in ocean currents, with negative impacts on fisheries, among other things.[5]
- Rising sea levels will flood populated coastal human settlements. They will also lead to more frequent and more intense storms and storm surges.
- The combinations of sea rise and desertification will severely undermine human economic and social networks.
- Increasing numbers of human beings will become climate refugees, suffering from increases in disease and pests, and fleeing impacted areas as they move to areas of the planet more capable of sustaining human life.
- This will inevitably lead to political instabilities and likely to increased violence.
- Approximately one quarter of the biological species currently living on earth will become extinct under the predicted conditions.[6]
- There will result a "radical impoverishment of ecosystems that sustain our civilization."[7]

A host of feelings arise spontaneously for anyone, lay person or scientist, who accepts these facts about climate change as true. Implicit in such feelings will be something like Lonergan's scale of values.[8] Projections of increases in human starvation and disease, the depletion of biological species, the desertification of land, and the loss of forests and bodies of water would evoke feelings about threats to vital values. Recognition of their devastating, large-scale impacts upon communities and their economic, social, and political patterns of human interactions stimulate feelings of alarm about the destruction of social and cultural values. The ensuing loss of human lives and the violence of war or conflict would give rise to feelings of horror at the annihilation of the revered value of human persons. In addition, the prospect of destruction of dynamic ecosystems of land, sea, and atmosphere that have been evolving for millions of years also elicits feelings of indignation among those who had come to value the even more encompassing emerging good that includes but goes beyond the scale of human values.[9]

The movement from learning the facts that climate science has revealed to feelings about those facts can be swift, overwhelming, and even paralyzing. Once the initial sense of being overwhelmed begins to subside, the structure of ethical intentionality can begin to operate. Feelings about the destruction of the vital, social, cultural, and personal values can lead to questions that move beyond feelings to judgments that objectively evaluate the state of climate change in all its complexity. Feelings and judgments of evaluation in turn lead spontaneously to further questions about what could and should be done. Here again computer modeling had something important to offer. As computer models became more sophisticated, they offered not only probable outcomes under current rates of GHG emissions but also probable outcomes under a variety of alternative parameters and altered circumstances. They offered alternatives that could be done. The scientists who knew about and understood the complexities of these computer models of alternative courses went on to form ethical judgments about what should be done to alter the dire projected outcomes. But they were not in a position to do what should be done by themselves alone. This could only be accomplished through the coordinated actions of a vast number of individuals and institutions. Educating and recruiting the public into this effort was, therefore, the necessary first ethical action.

By 1977 only a very small number of elite scientists even knew about these computer predictions and fewer still accepted them as true with a high degree of probability. It would be another decade before there would be a solid scientific consensus. Within that small circle of experts, a few took on the ethical responsibility to communicate what they knew to other scientists and to the wider public. Weart narrates what they did.

Raising Public Awareness

It has not been easy to bring facts about the realities of climate change before the public's attention, but that was a necessary ethical action for scientists to take. Initially, it was not even easy to raise awareness among scientists. This is because scientists tend to concentrate their focus within their own area of specialization, whereas the phenomena of climate change do not confine themselves neatly to one subspecialty within a science or even to a single science. Weart identifies this as a problem within the community of scientists itself, and especially for the scientific investigation of climate. He wrote that in the early 1970s: "Most scientists felt they were doing their jobs by pursuing research and publishing it. Anything important would presumably be noticed by science journalists and government science agencies."[10] This, however, was a mistaken assumption. Insights into *connections* among the many findings did not arise automatically, so there was a great need to develop ways of working in an interdisciplinary mode across scientific fields.[11]

In the previous chapter we saw that as scientists fashioned computer models of the climate, they inevitably had to learn how to collaborate with other scientists beyond their own specializations. They also had to respond to skepticism about whether what they were doing should even be counted as science. Responding to these challenges took both time and creativity.

The challenge is similar but magnified when it comes to communicating interdisciplinary research to public awareness. Public awareness is commonsense awareness. According to Lonergan, common sense involves the accumulation of innumerable insights, plus skills in focusing attention selectively, and practices of making good judgments, all about things of immediate practical concern. No one person possesses all the commonsense insights of a community,[12] but people also have commonsense insights about how to access the insights of others and how to communicate them to others when needed—*if* they belong to a well-functioning community. Ordinary commonsense people are practical; they are concerned with getting done the things that matter to them in the near future. Common sense has little concern with what lies far off. People of common sense develop skills of communication and mutual aid in order to collaborate with others who share their sets of practical concerns, but they are content to leave to others things not of their own concern[13]—which is not very different from what Weart observed about specialized scientists.

Raising public awareness as an ethical responsibility was taken on by a small number of scientists who recognized the importance of making the public aware. They made deliberate choices to take on that responsibility and gradually developed the kinds of tactics and skills necessary to be effective in that mission. Initially their efforts were only occasionally successful. At first only a few scientific findings regarding climate change ("global warming") made their way into the popular press. One remarkably prescient *Saturday Evening Post* article in 1950 predicted "a warmer planet; rising sea levels; shifts in agriculture; the retreat of the Greenland ice cap and other glaciers; changes in ocean fisheries, perhaps due to changes in the Gulf Stream; and the migration of millions of people displaced by climate change."[14] Most scientists then regarded the *Post*'s predictions as highly speculative, for at that time there was little if any scientific research to support these predictions.[15] In 1950, the few popular articles of this kind made very little impact.

After 1950 the scientific advances regarding global warming remained largely insulated from public opinion until the advocacy of Roger Revelle. Revelle was a well-trained scientist, having received his doctorate in geology from University of California at Berkeley in 1936. He had collaborated with Hans Seuss, making major contributions to climate science, but perhaps his most important role in the history of climate change was his effectiveness in communicating the scientific results to popular journalists and governmental policymakers.

Weart describes how Revelle took on this role early, even before the scientific community had reached a consensus about the most basic aspects of climate change. In 1956 and 1957, he testified before Congress, warning about the possibility of drastic environmental changes (although at the time he himself "did not expect much change in the climate for many decades, perhaps never"[16]). According to Fleming, Revelle's ability to access journalists and policymakers was enhanced by his connections through the prestigious Scripps family (he was married to family member Ellen Virginia Clark), and he worked at the Scripps Institution of Oceanography on and off for most of his career. He also worked in the Department of the Interior for the Kennedy administration, and in 1965 he presided over a major scientific conference, Causes of Climate Change. Fleming says that Revelle's connections gave him "godlike status" in the early public discussions of climate change.[17] Fleming reports that while Revelle did influence a few governmental policies, "the scientific work done in the mid-1950s did not seem to make much of an impression on the general public."[18]

Weart explains that Revelle's ability to gain even the level of public attention that he did achieve was prepared by unrelated postwar developments that heightened public awareness about the profound impacts that human technology could have on nature.[19] Fallout from the testing of nuclear weapons had become a great public concern. On a different front, Rachel Carson published *Silent Spring* in 1962, which gained attention because she was well known to the public through her highly regarded prior writings about the environments of the seas and shores.[20] Together, these and other such developments created a slight opening to theories about climate change.[21]

By the early 1970s, Reid Bryson joined Revell's efforts. Earlier Bryson had led major studies of the records of climate changes stretching over thousands of years. He concluded that abrupt changes had occurred in the past and were likely to occur in the future. Bryson became very vocal about the implications of climate research: "Scarcely any popular article on climate in the 1970s lacked a Bryson quote or at least a mention of his ideas."[22] Together, Revelle and Bryson were actively writing articles intended for a general audience, giving interviews and public lectures, and coming up "with quotable phrases for reporters. . . . Such efforts would reach, if not exactly the public, the small segment of the public that was well educated and interested in science."[23]

Revelle and Bryson were willing to make stronger predictions than science could support definitively at the time, "partly because the general mood of the times allowed it,"[24] although the public mood tended to shift easily back and forth among other immediate concerns. Attention did shift for a time toward global warming in the early 1970s by droughts and crop failures in the Soviet Union, India, Africa, and the American Midwest. Journalists flocked to Revelle, Bryson, and others for quotations. Once those catastrophes were past, however, public attention to global warming again waned. Unlike temporary droughts and crop failures, the complex dynamic of "global warming was invisible"[25] and not present to the immediate concerns of common sense as Weart observes, so a "wider public would take notice only if something special came along, something newsworthy."[26]

Lonergan would identify this difficulty as a "bias" specific to common sense itself: its "danger is to extend its legitimate concern for the concrete and the immediately practical into disregard of larger issues and indifference to long-term results."[27] So the effort to raise public awareness had to contend not only with the divide between scientific and commonsense modes of knowing but also with a prevailing "general bias" against any claim that goes beyond the immediate attention span of common sense.

Something special and newsworthy that could attract public attention did come along in 1977 when the United States was hit by the hottest July since the 1930s drought. This coincided with the NAS announcement about the dangers of global warming.[28] The simultaneous occurrence of the NAS declaration and the heat wave elicited a flurry of reporting in the popular media. By way of contrast, a scientific report would have attached less significance to this particular heat wave. After all, this one unusually hot July might be no more than a random deviation from an average mean for temperatures that itself had not changed. What counted for scientists was not *one* immediate, palpable event obvious to common sense, but long strings of data connected by theoretical and computer simulation advances into a complex pattern of rising *mean* temperatures. Different kinds of criteria were more compelling to scientists than convincing to the general public.

A year later the scientific conference in Vienna was still talking about global warming as a "possibility," which "was hardly news, and it caught little public or political attention."[29] Nevertheless, when pollsters asked in 1981 what people thought about increasing levels of CO_2 in the atmosphere, the results showed that "many people now suspected that they ought to be concerned about the greenhouse effect, but among the world's many problems it did not loom large."[30] The public had become somewhat aware, but the threat was not perceived as immediate or as calling for ethical and political action.

Over the years a number of unusual climate events and efforts by Revelle, Bryson, and others gradually contributed to a very modest shift in public awareness. Weart describes the situation:

> Up to this point global warming had been mostly below the threshold of public attention. The reports that the 1980s were the hottest years on record had barely made it into the inside pages of newspapers. A majority of people were not even aware of the problem. Those who had heard about global warming saw it mostly as something that the next generation might or might not need to worry about. Yet a shift of views had been prepared by the ozone hole, acid rain, and other atmospheric pollution stories, by a decade of agitation on these and many other environmental issues, and by the slow turn of scientific opinion toward strong concern about climate change. To ignite the worries, only a match was needed. This is often the case for matters of intellectual concern. No matter how much

> pressure builds up among concerned experts, some trigger is needed to produce an explosion of public attention.[31]

The igniting event, according to both Weart and Nathaniel Rich, was the summer of 1988, the hottest summer in US history to date. The heat wave of 1977 primarily impacted nations abroad, although it did affect the US Midwest. The heat wave of 1988, however, had a dramatic impact in the United States as a whole. Newspaper and television coverage again peaked, purveying "images of parched farmlands, sweltering cities, a 'super hurricane,' and the worst forest fires of the century."[32] Senator Timothy Wirt arranged for James Hansen to testify before Congress. "Hansen said it was time to 'stop waffling, and say that the evidence is pretty strong that the greenhouse effect is here.'" Shortly after Hansen's testimony, an above-average number of reporters traveled to Toronto to report on the World Conference on the Changing Atmosphere, which declared that the world's governments needed to set targets for reducing greenhouse gas emissions. "The story was no longer a scientific abstraction . . . it was about a *present* danger to everyone." A year later pollsters reported that 79 percent of US residents had heard or read about the greenhouse effect. This was an increase from 38 percent just eight years earlier.[33]

Significantly, this was also the moment when environmentalists took up climate change as a central "green" cause. Until this point in time, environmentalists had other reasons for preserving tropical forests, promoting energy conservation, slowing population growth, or reducing air pollution. While the environmental ethical issues discussed in part 1 did not disappear, they now became integrated into a larger, global concern for a whole that incorporated them. The laborious work of computer modelers had been gradually incorporating ever more components in order to arrive at increasingly accurate depictions of climate dynamics. To the initial crude models of the atmosphere as an undifferentiated whole were gradually added the environmental factors that made it into a more complex, layered reality. The interactions between ocean and atmospheric dynamics (such as calcification and acidification) were incorporated. These syntheses were followed by the inclusion first of more realistic differences across the earth's topographies, then of living ecosystems, and finally projections of human population growth and economic development. It is unlikely that computer models will ever capture all the rich complexity of the dynamic relationships among the earth's constituents (something most of the computer modelers themselves always admitted). Nevertheless, their efforts to find an integral structure that took seriously all the various components both gave them credibility and pointed in the direction of the need for a genuine integral structure—the heuristic structure of the emerging good—for facing the challenges posed by climate change.

A most important step in that direction came with the creation of the Intergovernmental Panel on Climate Change (IPCC) in 1988 by the World Meteorological Organization in cooperation with other United Nations agencies. From the point of view of public awareness and policy formation, this was one of the most insightful innovations in the history of climate change science and policy. Weart calls it a "unique hybrid" since it was "neither a strictly scientific nor a strictly political body."[34]

The IPCC was commissioned to issue an Assessment Report about every five years (there have been six so far). Each report was to be accompanied by the all-important "Summary for Policy Makers," in addition to other special reports that have been issued more frequently. The panel itself was composed of international governmental representatives who had close working relationships with scientists, but its work was carried out by hundreds of scientists and governmental officials working in task forces and committees, reviewing and debating the latest research publications and drafting reports. The reports themselves were debated and went through several stages of revision. This was a remarkable institutionalization of the self-correcting structures of cognition and ethical intentionality discussed in chapters 1 and 7. Weart credits the diplomatic skill and personal commitment of Swedish climatologist Bert Bolin for making the process work successfully. The reports had to be approved unanimously by the international members, so they were "negotiated word by word, were highly qualified and cautious." The first Assessment Report in 1990 only concluded that the world had been warming and that it would take a decade *before scientists could say with confidence* whether this was due to human activities. It did, however, identify "economically sound" ways to start reducing risk. Although this was less than many activists would have preferred, the slow and careful process insured that everyone's further questions were dealt with satisfactorily. As a result, "when the IPCC finally announced its conclusions, every word had solid credibility."[35]

Computer modeling formed the heart of the IPCC process.[36] Advances in data collection and new questions and ideas across specialties were incorporated into increasingly sophisticated, nuanced, and accurate computer models. Even questions arising from policymakers were taken into account. The growth of public awareness, the practical questions posed by governmental policymakers, and even skeptical resistance against scientists' reports had stimulated further scientific research and computer modeling. Weart writes that the "views of scientists and the public evolved together." He cites the 1995 IPCC report as "a striking demonstration of how the IPCC process deliberately mingled science and policy issues until they could scarcely be disentangled."[37] Yet it took almost twenty years after 1977 to achieve this degree of integration of scientific and public thinking and did not come about simply or easily.

The more advanced computer models were run under a variety of different circumstances and parameters producing reliable forecasts of likely outcomes. It became increasingly clear both what trajectory the world was on, and what likely outcomes would result from changes in various parameters. The "IPCC reports had become the output of a great engine of interdisciplinary research—a social mechanism altogether novel in its scope, complexity, and significance." In its Third Assessment Report (2001), the IPCC would finally go on record stating bluntly that the world was getting warmer and that greenhouse gases were the cause. Along the way, they responded to many "objections from industry-oriented skeptics and persuaded even the most recalcitrant officials."[38]

If 1977 was a watershed year for scientists, 1988 was the watershed year for public awareness. But public attention again began to wane after the 1988 heat wave abated. Other concerns began to occupy public attention. Even though a majority of the world's population was now aware of the phenomena of climate change, the sense of urgency ebbed and flowed as particular spectacular catastrophes came and went: Hurricanes Katrina (2005) and Sandy (2012) and the intense wildfires in the US Northwest, Canada, and Maui in 2015–2023, for instance, reawakened widespread public concern.

Although public awareness of what climate scientists were reporting gradually increased over the course of three decades, that awareness fluctuated, so it did not automatically translate into policy action. There were still other more formidable obstacles to overcome in moving from growing scientific knowledge to collective ethical commitments and policy actions. These obstacles have now expanded into a crisis with the rise of climate science denial, initially in the United States, but then into other nations across the globe. The next chapter will examine this most critical dialectical moment in the history of climate change ethics: the turn against climate science itself during the presidential administration of George H. W. Bush.

Obstacles to Raising Public Awareness

Before turning to the rise of climate science denial and its policy implications, this section delves more deeply into the question: Why was it so difficult to raise and maintain public awareness about the facts and values pertaining to climate change? Here Lonergan's functional specialty of Dialectic has something to contribute. Dialectic is dedicated to identifying the deep sources of conflicts, and the conflict in question here is between the ethical judgments and recommendations of climate scientists on the one hand and the recalcitrance of public commitment on the other. Among the deep sources of conflict that Lonergan identified in his discussion of Dialectic, two in particular are relevant to the present discussion: differentiations of consciousness and the already mentioned general bias.

The Problem of Differentiations of Consciousness

According to Lonergan's analysis most human thinking is commonsense thinking. As mentioned earlier, common sense is the vast accumulation of insights over many years. No one person possesses all the commonsense insights of a community; they are "parceled out among many" though shared with facility in a well-functioning community. Common sense is focused on what is concrete and palpable, on what will make an immediate difference to the interests and concerns of those in the community. Commonsense is concerned with getting things done in the near future.

On the other hand, the realm of scientific understanding (which Lonergan calls "explanatory" understanding) is concerned with how things are related to one another in general, not just how they are immediately related to the interests and concerns of a particular commonsense community.[39] As such, scientists are more detached from the limitations that common sense imposes on the kinds of questions and insights that matter. It has been said as far back as Plato that scientists and philosophers are concerned with truth, while ordinary, practical people are not. Lonergan has argued that this is not entirely correct. People of common sense are very much concerned with attaining truths about things that matter to them. The difference is not truth but the range of truths. The interests of scientists extend far beyond the public horizon. Scientists are concerned to understand truly a much wider range of phenomena that are seldom of concern to people of common sense.[40] Back when climate scientists were trying to determine what would happen if CO_2 levels doubled in a century or two, they were trying to attain explanatory understanding of how present conditions would be related to different conditions far off in the future. Such scientific concerns were not (yet) shared by people of common sense. As Weart point out, "Global warming was invisible, no more than a possibility, and not even a current possibility. . . . The prediction was based on complex reasoning and data that only a scientist could understand."[41] So there was and still remains a considerable problem of translating the evolving scientific knowledge into public understanding.

In principle, the sets of relationships explored by scientists can and should include how things are related to the interests and concerns of members of the commonsense community. In order to specialize and attain expertise in their complex investigations, however, scientists frequently become distant from everyday concerns, which has led to certain caricatures of scientists as absentminded and out of touch with practical realities.

Lonergan refers to this difference between commonsense and scientific modes of knowledge as a "differentiation of consciousness."[42] By this he means that people of common sense organize the native structures of their cognitional and ethical intentionality in ways that diverge dramatically from the ways that scientists organize theirs. The divergences of such differentiations of consciousness pose considerable problems for communication. These problems are not easily resolved and can lead on both sides to misunderstandings, conflicts, suspicions, hostilities, and animosities. Lonergan's method of Dialectic identifies differentiations of consciousness as important sources of conflicts, and he offers guidance for reaching reconciliation.[43]

What Lonergan called fully differentiated consciousness can move with ease between the two differentiations. But scientists no less than people of common sense tend to be more expert and at home in their own differentiation and less competent in the other. In order to communicate the findings of climate change science effectively to the public, scientists would have to think like people of common sense with facility. Lonergan held that ultimately this could only be accomplished completely by the development of a third differentiation of consciousness, which he called "interiority."[44] Interiority has a lengthy history in philosophy and literature, with figures like René Descartes and Michel de Montaigne playing seminal roles. It involves the development of techniques for attending to and accurately understanding movements within human consciousness. "It is only through the long and confused twilight of philosophic initiation that one can find one's way into interiority and achieve through self-appropriation a basis, a foundation, that is distinct from common sense and theory, that acknowledges their disparateness, that accounts for both and critically grounds them both."[45]

Ultimately, then, Lonergan regarded his own work, his basic method of self-appropriation, as the needed development. It would be unreasonable to expect every person of common sense to learn how to code at the most advanced level the computer climate models,[46] and equally unreasonable to expect every scientist to become adept in the countless numbers of commonsense cultures across the globe. The differentiation of interiority brought about by self-appropriation can recognize the movements of the self-correcting structures and realms of validity operating in both commonsense and scientific differentiations. On that basis it can mediate between these two differentiations and relate their different achievements to one another. This was the point of Lonergan's elaboration of his method of self-appropriation into a method of eight functional specialties. The objective of Lonergan's method is not to replace the ordinary ethical thinking of common sense and certainly not to replace the methodical work of science. Its objective, rather, is to offer timely interventions that help overcome counterpositions, misunderstandings, and conflicts that interfere with the normative developments in both commonsense ethics and science, so that they can cooperate and advance together to become increasingly ethical.

General Bias

The differentiation of commonsense from scientific consciousness presents an obstacle and problem calling for a solution. It cannot be solved easily, but the interiority differentiation and its methods does have the capacity to make constructive contributions. Yet there is also a more serious problem that Lonergan called "general bias," which is the tendency to extend the competency of common sense in matters "concrete and immediately practical into disregard of larger issues and indifference to long-term results."[47] Climate change is a long-term phenomenon that has been brewing since about 1760. No one has lived long enough to apply their own common sense firsthand in order to comprehend this centuries-long dynamic. Knowledge of climate change is based instead upon scientific extrapolations from past and present data toward future likelihoods.

It is one thing for commonsense people to honestly admit that they simply do not know about such things. It is quite a different matter to insist that such things do not matter or cannot be known. That is not ignorance but bias. The general mechanism of what Lonergan labeled as biases is their interference with the spontaneous self-correcting search for correct understanding in response to questions. Biases dismiss the questions or use other strategies to block the search for insights and judgments. Biases undercut the intellectual openness to learning what others have to teach us.

When people are subject to general bias, it is not only the quest for correct understanding that is undermined. Trust is also undermined. Anyone claiming to have a kind of knowledge about things that go beyond immediate facts and practical matters will be distrusted simply because of that claim. Trust in scientists was essential to raising public awareness and general bias tended to undercut such trust.

Trust and Belief

Even more significant than the challenges of mediation faced by interiority and its methods is the problem of trust. People of common sense do not know what scientists know because they are operating within different differentiations of consciousness. If people of common sense are to engage in serious ethical reflection about the facts that science has revealed, they have to *believe* as true what scientists say that they know. After all, it is only because scientists have come to regard the findings of climate science as true with a high degree of probability that they themselves respond with strong value-feelings and ethical judgments about what should be done.

Like the scientific community, the public, too, could only move toward responsible ethical courses of action on the basis of what it accepted as true about the climate situation. The public would have to believe what scientists said on the basis of their trust in them. Yet most scientists overlook the necessity and the challenges of fostering understanding, trust, and belief on the part of the public. This is because scientists also tend to underestimate the important roles of trust and belief in their own scientific practices. While people of common sense are prone to general bias, scientists, philosophers, and others can be afflicted by what Lonergan called "extra-scientific opinions."[48] These are counterpositions about the nature of scientific knowledge that are at variance with the actual cognitional structures utilized in the doing of science. One such extra-scientific opinion is that the facts are out there in plain sight for everyone to see, and if commonsense people do not see them, it must be because they are being deliberately obtuse. But scientific facts are not out there in plain sight to be seen. Again, as Weart astutely observed: "Global warming was invisible, no more than a possibility, and not even a current possibility. . . . The prediction was based on complex reasoning and data that only a scientist could understand."[49] What is out there to be observed are not facts but visual data. The facts came to be known not by taking a look, but through a historical, century-long self-correcting process in which hundreds if not thousands of scientists added their insights to the data and to the insights of others who came before them. Moreover, not even the data is out there for everyone to see. Only a very few scientists actually see the raw data from any given experiment. The rest of the scientific community trusts their reports. This trust is neither naïve nor blind; it is a trust that rests upon complex institutional structures that vetted the reports as believable and reliable. Trust and belief in reported results underpin the scientific community's knowledge of the facts of climate change, and therefore their feelings and judgments of ethical value.

Because they overlook the extent of trust and belief in their own disciplines, scientists therefore tend to underestimate the importance of building trust in order to overcome the barriers raised by differentiations of consciousness and to communicate what they have discovered to people of common sense. If scientists respond with condescension or disdain to the incomprehension on the part of commonsense people, this arouses and intensifies the general bias and undermines the building of trust. We will return to the issue of belief and trust in science and public responses in chapter 17.

Although raising public awareness about the reality and ethical issues of climate change had to contend with many difficulties, by 1988 much had been accomplished. In large part this was because of a general sense of trust of scientists on the part of the public. But the erosion and undermining of this trust was deliberately undertaken the very next year, resulting in a serious crisis in climate change ethics that remains with us today. This is the subject of the next chapter.

Chapter 15

The Dialectic of Politics and Climate Change Science

The rise in public awareness has gradually brought about some positive results. The ethical challenges set before the public by climate change have been met with countless insights into practical courses of action. Millions of individuals worldwide have acted on these ideas, from household energy conservation to more energy-efficient modes of personal transportation to planting trees and other CO_2 absorbing vegetation. Yet individual and residential consumption of electricity in the US is only 37 percent of the total electricity use, and direct emission of GHGs by residences is less than 12 percent of the total.[1] While individual efforts are important, the challenges can only be met effectively by coordinated collective actions.

Important collective actions can and, in a few cases, actually have been undertaken by corporations. Commercial, industrial, and agricultural businesses use 62 percent of the electricity and produce more than two-thirds of the GHG emissions in the US. Weart mentions that Walmart and General Electric were early leaders in making changes to address global warming in 2006.[2] Some other corporations have followed suit since then. Nevertheless, activists have complained about "greenwashing," the corporate practices of marketing themselves and their products as "eco-friendly" and "going green" without having made any real changes at all—they just rebrand their products. Others have made some changes, not because they were really serious about the climate, but rather because marketing themselves as "green" was good for market share. Rich explains that although Exxon began internal studies of global warming in 1979, it only did so because it "wanted to know how much of the warming could be blamed on Exxon."[3] Exxon temporarily became more serious in 1982, but that did not last.[4] There is much truth to the criticisms of the corporate responses and nonresponses to the findings of climate change science. Corporations still have a long way to go in meeting their ethical responsibilities regarding climate change. Nevertheless, wherever real reductions in GHG emissions have been made in products and modes of production, they *are* real contributions, though far from enough.

Should corporations be criticized when they make environmentally responsible decisions for the wrong reasons? Jonathan Sacks explains that in Jewish tradition, there are eight levels of *tzedakah* (justice + charity). The highest is giving aid to a person in need in such a way that the recipient regains dignity not only by becoming self-supporting but also becoming capable of rendering *tzedakah* to others. The lowest level, on the other hand, is meeting the legal obligation of Torah to render aid but doing so ungraciously. Nevertheless, Sacks points out that even the lowest level is counted as an act of *tzedakah* in Judaism because it does keep the law.[5] I think something similar can be applied to corporate actions regarding GHG reductions. So long as they really do take measures that reduce GHG emissions and do not do so in ways that increase GHG emissions elsewhere, their actions deserve to be called ethical, even if they are not done with the same spirit as is found in more genuinely dedicated corporations. It is certainly true that more dedicated corporations will be proactive in seeking ever better methods, but even those acting merely to meet legal requirements are part of the solution. *This means that law does play an important role in environmental and climate change ethics*. In this chapter, therefore, I will focus primarily on the history of governmental collective action, that is, the enacting of laws and the formation of administrative policies with regard to climate change.

The importance of law is not limited to its role in forcing corporations, other institutions, and individuals to take ethical actions in response to climate change science. There is also an important reciprocal relationship between governmental actions and public opinion. On the one hand, governmental action does shape public opinion. On the other hand, public opinion influences governmental action, especially in democratic societies. For example, Alliance 90/The Greens is now the fourth largest political party in Germany and has important influence on legislation. On the other hand, the leadership of Chancellor Angela

Merkel helped to strengthen the public commitment and support for GHG reduction policies in Germany. In the United States, Exxon once actively planned steps it could take to reduce GHG emissions, but it decided to wait upon governmental action (it did not want to be the only company making sacrifices).[6] Especially in democratic societies, government actions can influence public opinion, and the voters, ultimately, decide who will make laws and set government policies. Hence, the oscillating, dialectical political response to climate change science in the US has had significant ramifications beyond government actions themselves.

The Political Turn against Climate Science in the United States

It is not possible in this book to examine the history of governmental actions in all the nations that are members of the IPCC, let alone the actions of those that are not. I will therefore limit myself to the history of climate change policy at the level of the United States federal government. I have two reasons for choosing to focus on the US. First, the path followed by US politics brings to the surface several of the ways that biases of various kinds have interfered with achieving ethical outcomes in the area of climate change. Second, the path taken by the US gave birth to an international network of climate change denial that has had influences far beyond the US itself.

Although both Fleming and Weart mention how conservative political forces in the US resisted calls to GHG reductions, neither goes into much detail. In this chapter, therefore, I will rely on the historical study that Nathaniel Rich provides in *Losing Earth: A Recent History*.[7] Rich's narrative seeks to answer the question "How did the 'stubborn commitment to denialism' of climate science arise within Republican Party?"[8] Historically the Republican Party had been committed to environmental conservation and protection. In fact, Republican president Theodore Roosevelt has been called "the conservation president": he was one of the prime movers in both environmental conservation and preservation in the early in the twentieth century; he was one of the founders of the US Forest Service and designated five new areas as national parks during his presidency, more than doubling the total the number at that time; he also advocated the passage of the Antiquities

Act of 1906, which gave him and successive presidents the power to protect areas as national monuments. In the early 1970s, Republicans joined with Democrats in bipartisanship, passing the Clean Air Act (1970) and the Clean Water Act (1970), both during the presidency of Richard Nixon, also a Republican. By 1978, however, only a handful of Republicans voted in favor of the National Climate Act.

Still, in the early 1980s many prominent Republicans were concerned about the scientific conclusions and projections about the perils of climate change, and they were prepared to support legislation to implement solutions.[9] It was during the presidencies of Ronald Reagan (1981–1989) and George H. W. Bush (1989–1993) that the Republican Party shifted decisively and became a powerful opponent to governmental actions to mitigate GHG emissions. Rich tells the story of how this shift came about.

Rich focuses on two central figures who went to heroic lengths in advocating for governmental policies that would mitigate global warming: Rafe Pomerance and James Hansen. He begins his history in the spring of 1979, when Pomerance was Deputy Legislative Director at Friends of the Earth, an activist environmental organization. He later became Deputy Assistant Secretary of State for Environment and Development in the administration of President William Jefferson Clinton. Pomerance came across a paragraph in the lengthy "Environmental Assessment of Coal Liquefaction: Annual Report," published a year earlier by the Environmental Protection Agency (EPA), that said "the continued use of fossil fuels might, within two or three decades, bring about 'significant damaging' changes to the global atmosphere."[10] In his role at the Friends of the Earth, Pomerance had been focusing on air quality and air pollution issues, and he was "one of the most connected environmental activists." Yet even he had never heard anything like this before.

Pomerance learned that the pronouncement was also unknown to almost everyone in his circle of contacts, but he soon came to know Gordon MacDonald. MacDonald was a geophysicist who had a long career as a governmental advisor on security intelligence. In that capacity he became concerned about the possibilities of the US or its enemies "weaponizing weather," so he was already alarmed about impacts on the climate in the 1960s. Working with an intelligence advisory group, he contributed to a 1978 report to the Department of Energy entitled "The Long-Term Impact of Atmospheric Carbon Dioxide on Climate." The report warned of a possible increase in average global temperature of 2°C to 3°C, the melting of the ice sheets, a rise in sea levels, and the dire consequences for agriculture, drinking water, and their impacts upon human migration.[11]

That an environmental activist as well connected as Pomerance had not previously heard of MacDonald or the report reveals a great deal about the low level of public awareness in 1979. After being educated by MacDonald, Pomerance allied with him, and they began to make every governmental official they could familiar with the scientific predictions of the report. The story of what happened from this point onward well illustrates Weart's point about the mingling of "science and policy issues until they could scarcely be disentangled." In this case, scientific findings prompted political action, which in turn resulted in refinements of the scientific findings. After many weeks of listening to Pomerance and MacDonald's lobbying, the top science advisor in the Jimmy Carter administration directed the National Academy of Sciences, under the leadership of Jule Charney, to form a working group of scientists to determine whether their warnings were accurate. Charney's group studied the work of both Suki Manabe and James Hansen, leading to Hansen's recruitment as an advisor to the team. After long meetings, the Charney group concluded that Manabe had overestimated the amount of sunlight that would be reflected by snow and ice, while Hansen had underestimated it (based on their respective estimates of melting). As a result, the most likely rise in temperature for a doubling of CO_2 would be 3°C, and they issued their report.[12] Thus a high-profile governmental working group had concluded in July 1979 that the alarming implications proclaimed by Pomerance and MacDonald were realistic.[13] It is noteworthy that the working group was engaged in what Lonergan called the "processes of reflection to determine if there was a virtually unconditioned basis" for the judgments of fact that formed the conclusion of their report. In other words, this was not yet primarily a matter of scientists convincing elected officials (although that was a factor) but primarily a matter of convincing themselves that their own further pertinent questions had been properly answered.

The Carter administration was open to considering policies that would respond to the Charney report. The following year (1980), congressional hearings on global warming were held, and President Carter signed into law the Energy Security Act. It only authorized and funded another multiyear comprehensive study by the NAS, but the team included the best scientists and computer modelers in the US, including

Revelle, Manabe, and several prominent economists. Rich narrates how, as this committee was beginning its work, Pomerance spent a frustrating year trying to get scientific experts to agree upon politically compelling language and specific policy recommendations. At the end of that year, however, Carter lost the presidential election to Ronald Reagan, and the political tides changed dramatically: "Reagan floated plans to close the Energy Department, increase coal production on federal land, and deregulate surface coal mining. He appointed James Watt, the president of a legal firm that fought to open public lands to mining and drilling, to run the Interior Department. . . . After some debate about whether to terminate the EPA, Reagan relented and did the next best thing, appointing Anne Gorsuch, an anti-regulation zealot, who proceeded to cut the agency's staff and budget by a quarter."[14] Even though both Watt and Gorsuch were forced to resign in 1982, no governmental action was taken in response the Charney report. Nevertheless scientific studies continued to build an even stronger case.

Pomerance pondered how best to move the policy issues forward, and he decided that testimony before Congress by James Hansen would make a big impact. He liked Hansen's "blunt way of making the complex contingencies of atmospheric science sound simple" and believed that Hansen's "midwestern sincerity" would play well on Capitol Hill.[15] At almost exactly the same time, Representative Al Gore from Tennessee was chairing a subcommittee of the House Committee on Science and Technology. Gore had taken a class with Revelle while an undergraduate at Harvard, and he had been seriously committed to the climate issue ever since. Gore brought several experts to testify at his hearings, but he had little success until Hansen. During Hansen's testimony, Gore realized that the only question of concern to those present was "How much time until the worst begins?" Hansen ventured outside of his scientific expertise to communicate in commonsense political language: "Within ten or twenty years." A committee member then asked how soon they would need to change the national model of energy production. Hansen replied, "Very soon." Upping the ante on Hansen's reply, another scientist present, Nobel prize–winning chemist Melvin Calvin blurted out, "My opinion is that [the time] is past."[16] Both scientists' testimonies made the *CBS Evening News*.

The Reagan administration, however, resisted all such pressures, reserving its decisions until the completion of the three-year NAS study created by the Energy Security Act. In doing so, however, Reagan gave even greater prominence to the report. As a result, it would come to play a dominant role in US politics and policy formation for decades to come.

Thus another important turning point came in October 1983 when the NAS issued its report, *Changing Climate*. As a scientific document of 496 pages, it contained chapters and subsections not only on atmosphere, temperature, and seas but also agriculture, economics, and social policy. It contained all the technical details, data, and nuanced conclusions that are expected of scientific method. It explored the prospects of frightening possibilities, such as "an ice-free Arctic . . . Boston sinking into its harbor . . . [and] political revolution." In the preface, the committee's chairperson, William Nierenberg, argued that "action had to be taken immediately *before all the details could be known with certainty*, or else it would be too late."[17]

That, however, was *not* what Nierenberg or other committee members said in the press interviews that followed the release of the report. "They argued the opposite: there was no urgent need for action. Nierenberg warned that the public should not entertain the most 'extreme speculations' about climate change."[18] This was because *absolute certainty* about those predictions was not yet there, which led to dialectical conflicts even among scientists themselves. The same week as the NAS report was made public, the Environmental Protection Agency (EPA) also released a two-hundred-page report with far more dire predictions and much greater urgency for immediate policy initiatives. But one prominent member of the NAS study committee, Joseph Smagorinsky, a pioneering climate modeler, "openly denigrated the 'unnecessarily alarmist' EPA report." Nierenberg called that report "a badly done thing," and other members of the committee joined in this chorus.[19] The *New York Times* ran a front-page story with the headline "Haste of Global Warming Trend Opposed," which reported that "the coming warming of the earth caused by a buildup of carbon dioxide in the atmosphere is 'cause for concern' but that there is sufficient time to prepare for its impact." It quoted Reagan's top science advisor, George A. Keyworth, who condemned the EPA report as "unwarranted and unnecessarily alarmist."[20] The spokespersons for *Changing Climate* were prominent scientists; they had spent three years working on this report; and they were unified in their public statements about the degree of *scientific certainty and uncertainty* about future predictions and the degree of urgency. Almost everyone believed them rather than the EPA report. The American Petroleum Institute and Exxon both quickly terminated their task forces for considering alternative production policies. Any hope for major US policies to counter the buildup of CO_2 was dead in the water.

Pomerance was outraged. In desperation he convinced Gore to hold more hearings, trying to overcome the damage done by the way *Changing Climate* was being portrayed. While he, Wallace Broecker, Carl Sagan, and others tried to make the case for a more urgent situation, others (notably NAS member Thomas Malone) made the counterargument. The hearings did not change the situation. Pomerance withdrew from the fray for a year to ponder what to do next.

Weart's remark that "no matter how much pressure builds up among concerned scientists, some trigger is needed to produce an explosion of public attention"[21] again describes this situation. For the Reagan administration's stance toward climate change, that igniting event came in May of 1985. British researchers had documented a shocking depletion in the ozone layer over Antarctica—what came to be known popularly as "the hole in the ozone layer." Ozone filters out the highest frequencies (shortest wavelengths) of ultraviolet radiation (UVC) from the sun. UVC causes significant increases in skin cancer and is destructive to agriculture and fish stocks.[22] Alarming press and television reports exploded around the world. A time-lapse video of satellite images created by Robert Watson depicting the growth of the "hole" was shown globally.[23] In response, President Reagan proposed a 95 percent reduction in CFC production, and he submitted to the Senate for ratification the UN global treaty "World Plan of Action on the Ozone Layer," which until then had been stalled. This "Montreal Protocol" was signed by more than three dozen nations in September 1987.[24]

Rich explains that ozone depletion became conflated in the public imagination with GHG-induced global warming. This was partly due to Hansen's research on the ways that CFCs function as GHGs, although the chemistry and physics were very different. Nevertheless, "the confusion helped."[25] Global warming and GHG emissions were not only back in the public awareness but now carried an intensified sense of political urgency. Several months after the announcement of the ozone depletion, eighty-nine international climate scientists gathered in Villach, Austria, to reconsider the question of the urgency of limitations on GHG emissions. Malone, who had a year earlier used the findings of *Changing Climate* to undercut the EPA warning, underwent a "notable conversion" and announced: "I believe it is time to start on the long, tedious, and sensitive task of framing a convention on greenhouse gases, climate change, and energy." Rich observes, "The formal report ratified at Villach contained the most forceful warnings yet issued by a scientific body."[26] Due to the widely viewed impact of a time-lapse video showing the growth of the ozone hole, Pomerance realized that the public had become alarmed because "ordinary people could be made to see" the ozone hole grow.[27] Something similar was needed to further advance the case for global warming. As Lonergan would put it, it was necessary to address the general bias of common sense that privileges what is immediate and present, as well as the counterposition that one has to see with one's own eyes to know anything. Modern science, on the other hand, did not need to overcome such a prejudice because it had developed a differentiation of consciousness that could intelligently and reasonably extrapolate from present data to future consequences.[28]

Climate issues were once again making headline news and receiving increased attention at congressional hearings. There was widespread bipartisan support to work on climate policies. Pomerance and MacDonald decided to have Hansen testify at one high-profile hearing. By now Hansen was working under the National Aeronautics and Space Administration (NASA) and it was a matter of course for a federal employee to submit his testimony for approval beforehand. He did not anticipate any difficulties, but he was unexpectedly directed to change his testimony. The White House objected to at least two of his sentences: "Major greenhouse climate changes are a certainty" and "By the 2010s [in every scenario], essentially the entire globe has very substantial warming." No reasons for these objections were provided.[29] Hansen refused to change his statements, which meant that as a NASA employee, he would not be allowed to testify. A compromise was reached in which he would testify as a private citizen, but that robbed him of the authority of his position as a governmental researcher. He planned to say, if asked, that he was not testifying in his official role because the White House had insisted that he make false statements, but no one asked him about this. His hope to expose the White House witness-tampering was dashed. Hansen "realized that there were people at the highest levels of the federal government—within the White House itself—who hoped to prevent so much as an honest reckoning with the nature of the problem. This, it seemed, was a new development: not merely an expression of indifference or caution, but the emergence of an antagonistic—a nihilistic—force."[30]

Once again, however, nature provided an igniting spark to keep things moving politically. As mentioned in chapter 14, 1988 was the hottest summer on record. June 23, 1988, recorded the hottest temperature on record in Washington, DC, for that date. This provided a spotlight and drew attention to Hansen's powerful testimony on that exact day. Hansen concluded his testimony with his famous exhortation: "It is time to stop waffling . . . and say the greenhouse effect is here."[31] Once again the media were generating vivid news stories linking the hot summer to the science of global warming. "For the first time the crisis had a face and, with it, an emotion—a fever of pent-up frustration, outrage, and moral conviction."[32] Value feelings were spontaneously responding to reports accepted as factually true. The effort to bring about climate change policies was back on track.

Three days after Hansen's testimony, Pomerance realized that the movement needed something different from the technically exact and statements of carefully qualified certainty and uncertainty by scientists. It needed something that would address common sense. It needed a concrete, hard target, like the 50 percent reduction of CFCs built into the Montreal Protocol. Pomerance came up with "20% by 2000," a slogan that had commonsense effectiveness. There was no exact scientific support for this number or date, but it was the sort of thing that could galvanize public support. He consulted with numerous scientists and environmental officials, who all agreed this was at least in the ballpark. He announced this target in his keynote address at an international conference in Toronto, and all four hundred scientists and politicians in attendance signed on to 20 percent by 2005.[33]

The heightened concern for climate affected the 1988 presidential campaign. Candidate George H. W. Bush declared, "I am an environmentalist," and he made campaign promises to address the environment. By the end of the year thirty-two climate bills had been introduced into Congress, including one that would commit to a 20 percent reduction in GHG by 2005. The IPCC had been created the same year. As an initial show of his commitment to climate and environmental issues, newly elected President Bush invited the working group of the IPCC to convene in Washington, and it met at the State Department ten days after his inauguration.[34] Brand-new secretary of state James Baker addressed the meeting, saying, "We can probably not afford to wait until all of the uncertainties about global climate change have been resolved before we act."[35]

This was exactly the position Hansen, Pomerance, and many others had been advocating in support of the *Changing Climate* report. Scientists as scientists could not claim absolute certainty about their facts and their projections. As scientists, they recognized further questions that still needed answers. Yet their integrity in openly acknowledging the lack of absolute certainty in their findings seemed to undermine the very pragmatic efforts they wanted to succeed. Achieving pragmatic, commonsense political success in policy change required rhetoric that went beyond the kinds of language many scientists were trained to use as scientists. This is an instance of what Lonergan calls the dialectical conflict that results from the differentiation of scientific from commonsense consciousness. As scientists, they acknowledged some degree of uncertainty in science, but, as advocates of change, they needed a different way of communicating to people of common sense and to elected officials. Baker, Hansen, Pomerance, and others were attempting to bridge this gap in communication. They addressed scientific uncertainties the way people of common sense do when they make decisions to buy insurance policies even though they lack absolute certainty about whether their homes, automobiles, or health will actually suffer catastrophes. Acting on probabilities rather than absolute certainties is exactly the right ethical position whether buying personal insurance or insuring the planet.

At that time, the petroleum industry itself was still open to debating the best courses of action in light of the scientific reports, at least among some influential players. In the fall of 1988, Terry Yosie of the American Petroleum Institute organized a conference attended by over nine hundred representatives from the petroleum industries. They heard from speakers that "the costs of controlling climate change were still fairly modest and worth pursuing"; that investing in alternative energy sources could bring profits as global heating increased demand for air conditioning and refrigeration; that "a 2-degree warming could stagger the global economy"; and that the "longer industry waited to act, the worse it would go for them." Yosie himself told the audience that "uncertainties remained, such as the timing of the changes. But the trend . . . was not in doubt." Though their openness was largely motivated by utilitarian self-interest—namely, how to avoid accusations and how to make profit from investing in alternatives—the petroleum industry at that time still regarded climate science as worthy of consideration. They even created a modest fund of $100,000 to further investigate the question, although the industry as a whole remained indecisive.[36]

Yet at the very same time a rising movement was working at cross-purposes to undercut the efforts of Hansen, Pomerance, and others. The opposing strategy was to emphasize the uncertainty of climate change science, and it was given a decisive thrust by President Bush's chief of staff, John Sununu. He vehemently disagreed with Secretary Baker's exclamation that "we can probably not afford to wait until all of the uncertainties about global climate change have been resolved." In the spring of 1989, Sununu told Baker, "You don't know what you are talking about," and he silenced Baker from speaking on climate issues thenceforth in his official role. More than any other single individual, Sununu was responsible for the alienation of the Republican Party from environmental protection and climate change mitigation policies, which has mired American politics in a seemingly irresolvable conflict ever since. While governor of New Hampshire, Sununu had rebuffed environmentalists' opposition to the construction of a nuclear power plant, even though he had previously supported legislation to end acid rain. He was educated as an engineer and served as an advisor at MIT's graduate program, yet "he harbored skepticism toward scientists who mingled" science and political advocacy. He believed that science had been used to advance anti-economic growth doctrines.[37] He became increasingly suspicious of Hansen's scientific findings as "technical garbage"[38] and had "a rudimentary one-dimensional general circulation model installed on his personal desktop computer." On this basis, he "decided that Hansen's models were horribly imprecise, 'technical poppycock' that failed adequately to account for the ocean's capacity to mitigate warming."[39] The model he used, however, was one of the earliest; it had been superseded not only by Hansen's own later work on ocean mitigation of warming but also by the much more sophisticated work of the many computer models of the climate that had been done since.

Apart from Sununu, there was initial openness in the Bush administration as well as in Congress about policies to cut GHG emissions. As we have seen, for a short time there was even serious consideration at the American Petroleum Institute about the costs and benefits of cutting emissions. Even though President Bush, several of his appointees, congresspeople, and some petroleum CEOs were considering policies for reducing emissions, there was still indecision. Eventually Sununu's power eclipsed every effort at substantial policy innovations. Two incidents are especially significant. One concerns the IPCC meeting in Geneva in 1989, and the other concerns James Hansen. The first incident came as a response to EPA administrator William Reilly's advocacy for the US to take the lead at the IPCC conference in advocating for GHG reduction targets. "Sununu disagreed. It would be foolish, he said, to let the nation stumble into a binding agreement on questionable scientific merits, especially one that would compel some unknown quantity of economic pain."[40] He ordered Reilly and the other US delegates to make no commitment to the IPCC proposals presented in Geneva.

The second incident took place in April 1989. Senator Gore again called Hansen to testify in hopes of moving the president to take a more decisive stand. Hansen again had to submit his testimony to his superiors for review. Once again, the administration (Sununu this time) demanded changes. This was the second time that Hansen had been told to change his testimony in ways that he knew were contrary to his scientific knowledge. He had had enough. He contacted Gore and told him what had happened, and they set a trap. When Hansen testified, Gore pretended to be surprised, and asked why Hansen had contradicted his previous statements and publications. Hansen explained that his original statement had been altered by the White House, remarking, "My only objection is being forced to alter the science." Another government scientist who was also present at the hearings said that his testimony had also been altered. Gore became indignant, the press had a field day, and later that day the president's press secretary admitted that they had been required to alter their statements.[41] As a result of the pressure, Sununu contacted the US delegation in Geneva and grudgingly reversed his order, now saying that they should "work to develop full international consensus" toward a treaty process.[42]

For a few months the Bush administration and even Sununu again seemed to be serious about the climate change crisis. At the next meeting of the IPCC in November 1989, however, Sununu ordered the US delegate to abandon any commitment to freezing GHG emissions at then-present levels, and he lobbied Great Britain, Japan, and the Soviet Union to agree. On the brink of the first international agreement to curb GHG emissions, Sununu sabotaged the US' chance to exert positive leadership.

This was the decisive turning point in the history of climate change politics. Although Sununu eventually resigned due to improprieties, by the time he did so the President's Council of Economic Advisers had turned against sacrificing economic growth for climate security. Although the US did sign the Rio Accord (except for the goals on preserving biodiversity), it played more of an adversarial than a leadership role at the IPCC meeting in Rio de Janeiro in 1992 (the "Earth Summit"). When

George H. W. Bush was defeated by Bill Clinton with Al Gore as his vice president, it seemed that climate change policy momentum had again gained the upper hand. Even though the US delegation (which now included Pomerance) to the next IPCC assembly in Kyoto, Japan, in 1997 endorsed the protocol proposed there, the US Senate voted 95–0 peremptorily against its major component, so it was never ratified. While the political reasons for Clinton's failure to use his influence in this case are complicated, the lack of trust in the certainty of the science of climate change was a key factor.

The United States did become a signatory to the Paris Agreement in April 2016, and President Barack Obama accepted it by executive order in September 2016. The agreement was signed by 195 countries. That treaty committed all signatories to strategies intended to limit global warming to less than 1.5°C above preindustrial levels. Obama also committed the United States to contributing $3 billion to the Green Climate Fund.

President Obama's initiative was short-lived. Soon after his election, Republican president Donald Trump announced that the US would withdraw from the Paris Agreement of 2016, citing it as a "bad deal" for America. The Trump administration took many other steps to block or reverse initiatives that would have helped realize the goal of reducing GHG emissions, as well as numerous other environmental protections. Fortunately these destructive policies were reversed by the Biden administration beginning in January of 2021, although four precious years were lost in during the opposition under the Trump presidency.

It would be a mistake, however, to assume that the election of President Joseph Biden has resolved the crisis. Climate change denial and policy opposition did not begin with Donald Trump. He was elected president in part precisely because of his opposition to climate control policies, and even because of his publicly expressed doubts about the reality of climate change or the role of human causation in it. Trump appealed to a significant portion of the electorate that had been shaped by a widespread network of climate change denial, which had been growing well before the election of 2016.

The Growth of Organized Denial

The idea that climate science was "uncertain" has infected the American political process almost irrevocably. When Sununu used his powerful political position to publicly cast doubt on the certainty of climate science, he stimulated the exponential growth of a national and international network of climate change denial, as documented by Rich, Weart, and Riley Dunlap and Aaron McCright.[43] The climate denial movement gathered momentum from Sununu's initiative as political and business people who were originally open to considering the challenges of climate change science now shifted direction. Those who had long been opposed in principle to all forms of government regulation now seized upon this new strategy: rather than attack climate change policies because they were governmental interference in the free market, their new approach was to attack climate science itself on the basis of its uncertainty. Rich and Weart narrate the creation of organized climate denial during this period. In response to the formation of the IPCC, the speciously named Global Climate Coalition was formed in 1989 by petroleum and other corporations. Its stated mission was to become "the leading voice for industry on the global climate change issue." However, its real mission was to undermine confidence in climate science by emphasizing the uncertainty of scientific findings.[44]

Numerous other climate change denial organizations (many financed by the petroleum industry) began to proliferate. Their chief tactic has been to emphasize "the uncertainty in scientific conclusions."[45] Dunlap and McCright narrate the twenty-year growth of a worldwide network that adopted this strategy, a network they called a "well-funded, highly complex, and relatively coordinated 'denial machine.'"[46] This network comprises fossil fuel corporations, numerous institutes and associations funded by them (e.g., American Petroleum Institute), conservative philanthropists, foundations and think tanks (e.g., Citizens for a Sound Economy), front groups (e.g., Global Climate Coalition), contrarian scientists and their institutes (e.g., George C. Marshall Institute), media outlets (e.g., Fox News and the *Weekly Standard*), along with elected officials who regularly use the claims promulgated by these organization in their campaigns. The strength and coordination of this network had its beginnings in the US in the late 1980s but quickly took root and spread through Canada, Australia, Great Britain, and elsewhere. Today there are active climate denial networks in virtually every nation. They have gradually infused a stubborn hostility to climate change science into the electorate of virtually every nation.[47] In subsequent chapters we will return to the legitimate questions about economic development that must be faced by ethical reflection about climate change. But the deliberately orchestrated sowing of doubt about and denial of climate science remains alive to this day.

The Basic Dialectical Factors

The election of President Joseph Biden did enable the US to return to the Paris Agreement and to reverse many of the Trump administration's policies. These have been important events, but the cultural factors underlying climate change denial have not gone away. Those cultural factors have had a major influence on the "dialectic" of the politics of climate change, and as such they need to be addressed on many levels. The purpose of this section is to pave the way for a critical review of the foregoing history through the lens of Lonergan's dialectic in order to understand and resolve the most basic issues driving the political and public reversals on climate change.

The tendency of political debates and policies during this period to seesaw back and forth is an example of what Lonergan means in characterizing history as a "dialectical" process. For Lonergan, history is a process where two or more fundamental factors are in conflict with one another. As supporters of one factor implement their decisions, supporters of the other factor change their strategies in response. This process repeats over and over, each side gradually modifying itself in efforts to maintain their fundamental positions. While the tactics may change, the conflict will remain unresolved until the root sources are addressed. Identifying and resolving these basic sources was one of the major objectives of Lonergan's method of eight functional specialties, especially the method of Dialectic.[48]

The historical studies by Rich and Weart reveal two principal cultural factors that underlie much of the political conflict that arose in response to climate change science: scientific uncertainty and strong opposition to government interference in private and corporate decision making.

A type of uncertainty is in fact part and parcel of all modern science, especially when it comes to phenomena as complex and intertwined as those of climate change. In the next chapter, we will employ Lonergan's method of Dialectic and offer a much-needed philosophical clarification of the status of scientific knowledge as probable versus certain or uncertain.

The second conflict underlying the dialectic of politics and climate science arises out of the strident culture of opposition to government interference. It has its roots in the "deep stories" of American cultural values. In the next section, I will explore the ways that a number of

American cultural values have accelerated the growth of climate science denial and how this is traceable to a truncated notion of value.

The Deep Stories of Culture and the Great Paradox

In *Strangers in Their Own Land*, Arlie Russell Hochschild set out to find answers to what she called "the Great Paradox."[49] That paradox was why the people most severely in need of relief from a range of serious environmental problems were among those most vehemently opposed to governmental actions that would bring such relief. Her additional goal was to understand how this paradox affects American politics in general. A sociologist by profession, Hochschild conducted sixty interviews, attended gatherings, and got to know communities of people in regions of Louisiana that had been ruined by chemical and energy corporate practices. In her effort to understand this paradox, she chose the "keyhole issue" of pollution and destruction of the environment by corporate actions. She chose Louisiana in particular because of the severity of such impacts there and because of the state's extreme poverty, which left it with inadequate local resources to redress the environmental damage. Louisiana is the second poorest state in the nation, ranking forty-ninth out of fifty on an index of human development. Louisiana is also the fourth largest energy producing state, but it receives among the lowest tax payments from energy companies in comparison to other states.[50]

Based on her interviews, Hochschild discovered a phenomenon that other researchers found elsewhere in the US: "If, in 2010, you lived in a county with a higher exposure to toxic pollution, we discovered, you are more likely to believe that Americans 'worry too much' about the environment and to believe that the United States is doing 'more than enough' about it."[51] As she got to know the people more personally, she gradually came to understand that "the Great Paradox becomes more complicated than it first seemed."[52] As she demonstrated through reports of her numerous interviews, these were *not* people who had no concern for the environment. Rather, she documented in great detail that these people cared greatly about preserving the environment. They lamented loss of fishing and hunting spots that they had enjoyed in their youth. They had lost meaningful work and community friendships through callous disregard on the part of corporations. Some had actually lost their very homes to sinkholes that resulted from industrial dumping, and in many cases their health had been severely impacted. Even so, the vast majority of them supported political parties and candidates that opposed governmental regulation of the companies that had made their lives and environments so miserable.

She gradually identified two "deep stories" that constitute American culture, and she concluded that these deep stories have a greater influence on people's decision making than almost anything else. A "deep story," as she puts it, "is a feels-as-if story—it's the story feelings tell, in the language of symbols. It removes judgment. It removes fact. It tells us how things feel. . . . And I don't believe we understand anyone's politics, right or left, without it. For we all have a deep story."[53] Deep stories communicate by embedding symbols in songs, legends, jokes, stories, posters, plays, movies, advertisements, political cartoons, and so on. The symbolisms in deep stories evoke and reinforce mixtures of feelings.

Deep stories cultivate fundamental feelings for values that guide and shape cultures, which in turn guide and shape political decisions. Lonergan argued that these deep value commitments of a culture determine how people choose to organize themselves into the interlocking networks through which they collaborate to accomplish the goods and maintenance of their social order (what he called a "good of order").[54] Without having read Lonergan, Hochschild and other scholars have come to the same conclusion. She identified two deep and conflicting stories that suffuse American political institutions and decisions.

The first of the two deep stories Hochschild calls "Waiting in Line." It is arguably the dominant deep story in American culture. In order to gain a sense of the symbolic communication of feelings and values embedded in this first deep story, it is best to quote her at some length:

> You are patiently standing in a long line leading up a hill, as in a pilgrimage. You are situated in the middle of this line, along with others who are also white, older, Christian, and predominantly male, some with college degrees, some not.
>
> Just over the brow of the hill is the American Dream, the goal of everyone waiting in line. Many in the back of the line are people of color—poor, young and old, mainly without college degrees. It's scary to look back; there are so many behind you, and in principle you wish them well. Still, you've waited a long time, worked hard, and the line is barely moving. . . .

> You have shown moral character through trial by fire, and the American Dream of prosperity and security is a reward for all of this, showing who you have been and are—a badge of honor. . . .
>
> Look! You see people cutting in line ahead of you! You're following the rules. They aren't. As they cut in, it feels like you are being moved back. . . . Through affirmative action plans, pushed by the federal government, they are being given preference for places in colleges and universities, apprenticeships, jobs, welfare payments, and free lunches. . . . It's not fair . . .
>
> Unbelievably, standing ahead of you in line is a *brown pelican*, fluttering its long, oil-drenched wings. [Receiving governmental protection] it's in line ahead of you. But really, it's just an animal and you're a human being.
>
> Blacks, women, immigrants, refugees, brown pelicans—all have cut ahead of you in line. . . . You resent them, and you feel it's right that you do. So do your friends. Fox [network] commentators reflect your feelings, for your deep story is also the Fox News deep story.[55]

Hochschild shared her synopsis of this deep story with the people she came to know personally through her interviews, asking if it resonated with them. Overwhelmingly they endorsed this as their own deep story. One said, "*I live your analogy*." Hochschild comments: "For Louisianans I came to know, the deep story was a response to a real squeeze."[56] They believed that their wages had not kept pace with prices.

Later in the book she describes four central "representative characters," which, as she explains, anchor the symbolism of this deep story: the Team Player, the Worshipper, the Cowboy, and the Rebel. Regarding the symbolism of the Cowboy she writes: "The Cowboy expressed high moral virtue. Equating creativity with daring—the stuff of great explorers, inventors, generals, winners—Donny honored the capacity to take risk and face fear."[57] She describes a debate between Donny and his friend but antagonist, Mike, about the dumping of industrial toxic waste that had ruined the local clay beds and imperiled a local bridge along I-10. For Mike, government regulation is necessary to clean up the toxic effects and prevent them in the future. For Donny, every individual has to make individual choices and bear risks, so any government regulation is just another blow to individual freedom. Hochschild shows how this deep story draws upon the iconography of the Cowboy whose courage and cunning allowed him to rise above all risks in the Wild West.

Numerous scholars have filled out in great detail the historical development of the American cultural deep stories and iconic symbolic figures. These include W. E. B. Du Bois, Henry Nash Smith, Leo Marx, Michael Kammen, Garry Wills, Richard Slotkin, Andrew Delbanco, Isabel Wilkerson, Heather Cox Richardson, and Kristin Kobes Du Mez, among others.[58] They explain the long historical development of the American deep stories and especially the evolution of the central symbolic figure: the lone, self-reliant Frontiersman and then Cowboy (and now rogue Urban Detective or vigilante Superhero) who stands outside of urban legal society and all alone conquers all adversities, natural as well as human. They narrate how the deep story and its central symbolic figures developed gradually (and dialectically), beginning with the earliest English-background settlers who struggled to adapt to an alien terrain and climate of North America. As Wilkerson puts it, "We are all players on a stage that was built long before our ancestors arrived in this land. We are the latest cast in a long-running drama that premiered on this soil in the early seventeenth century."[59] The symbolisms of the deep story reinforce feelings of honor for being self-reliant as well as feelings of distrust and hostility toward cities, civilization, and government. These scholars also document how the deep story was communicated in popular literature over several centuries. Most importantly, the deep story has been passed along in ordinary everyday conversations and interactions that both draw upon and supply material for popular literature. These scholars also show how this deep story has been used both to perpetuate white supremacy in a succession of guises and to relegate women to an inferior status. They also unveil its ideological nature, its paradoxical effect of blinding people to their own exploitation by the growing centralization of corporate and financial power, to the detriment of what would actually be in their own best economic interests. Those who identify with this deep story have in fact been marginalized from attaining the American Dream by corporate decisions that have destroyed their environments and eliminated their jobs, the latter of which is one of their primary sources of self-worth. But their feelings have been cultivated to cast blame and resentment in the direction of the Line-Cutters and the government that finds them worthier.

The link between this deep story and opposition to the governmental role in environmental and climate protection is anchored for

Louisianans by the symbol of the Brown Pelican. The brown pelican is Louisiana's state bird. Hochschild narrates how the brown pelican became an endangered species, and how government regulations (e.g., concerning chemical pollution and hunting) brought it back out of endangered status. Just one year after the bird was removed from the endangered species list, however, the survival of the brown pelican was again threatened by the massive BP oil spill in 2010. Through the eyes of the deep story, the great expenditure of government time and money on protection of the brown pelican and other environmental elements came to be seen not only as stealing resources that could have helped the people of Louisiana advance toward the American Dream but also as a clear sign that they themselves are regarded as less worthy than the brown pelican. They felt that the government was favoring the brown pelicans (and the environment and climate in general) over the interests of hard-working common people, who were in turn pushed even further back from realizing the American Dream. Because of its prominence, the Brown Pelican became an anchoring symbol in the deep story in Louisiana. Elsewhere other things became symbolic of environment and climate Line-Cutting: Northern Spotted Owl, Snail Darter, Tree Huggers, and even Global Warming itself became a symbolic Line-Cutter. "A former Democrat, Donny voted for Republican presidential candidate George W. Bush because 'if Al Gore believes in climate change, he's too stupid to be president.' Since then, he had moved to the right of the Republican Party."[60]

Hochschild also describes a second, conflicting deep story, although she does not elaborate it in as much detail as the previous one. This is the deep story of those "on the left," with whom she expresses her own sympathies. The symbol of waiting in line to attain the American Dream over the hill still remains central in this second deep story, but this story focuses on those at the back of the line: "Blacks, women, immigrants, refugees," and also the environment. Unlike those ahead of them in line, these constituencies have continuously been denied the means to advance. Every effort they make to move forward in the line has been blocked by discrimination, injustice, and violence. They are "victims." Like the first deep story, this is also a story about what is fair and just, yet it is a story about making up for past injustices.

In the most recent rendition of this second deep story, the environment and climate themselves came to be viewed as victims (as in some proponents of deep ecology). Injustice in this deep story is not a matter of Line-Cutting but of Line-Obstruction. This second deep story has gradually begun to link its various elements under the themes of environmental racism and environmental justice. These themes draw attention to the fact that those who have long suffered from Line-Obstruction are also disproportionately impacted by environmental pollution and climate change. Believers in this story look upon the government as an agent with the obligation to make up for the past injuries that individual efforts alone cannot overcome because of unjust impediments. While those who live out this deep story are more likely to recognize corporations and financial institutions as responsible for the past injustices, there is, at least implicitly, also a tendency to feel antipathy toward those who subscribe to the other deep story as equally responsible for those past injustices. There are also hints of blaming adherents to the other deep story for environmental and climate harm, not only because of the politicians they support but even because of their very own lifestyles.

There are of course themes and variations on these two deep stories, which the scholars have elaborated in detail. There are also other deep stories that play less dominant (though not less important) roles in American culture, which have also been brought to attention in the growing body of scholarship. These two deep stories, however, have played the greatest roles in the formation of public policy in the United States regarding climate change. The Republican Party had long formed its political base by appealing to the first deep story to gain influence and elected office. In the first section in this chapter, we saw how John Sununu played the critical role in appealing to the first deep story and placing climate change science into that story as one of its villains. He thereby turned the Republican Party against climate change science and the need for ethical and policy responses to it. In doing so, he was both a believer in and an exploiter of the sentiments cultivated by the Line-Cutter deep story.

Regarding the dialectical conflict between these two stories, Hochschild writes:

> Given our different deep stories, left and right are focused on different conflicts and the respective ideas of unfairness linked to them. The left looks to the private sector, the 1 percent who are in the over-class, and the 99 percent among whom are an emerging under-class. This is the flashpoint for liberals. The right looks to the public sector as a service desk

> for a growing class of idle "takers." Robert Reich has argued that a more essential point of conflict is in yet a third location—between main street capitalism and global capitalism, between competitive and monopoly capitalism. "The major fault line in American politics," Reich predicts, "will shift from Democrat versus Republican to anti-establishment versus establishment." The line will divide those who "see the game as rigged and those who don't."[61]

Hochschild and Reich may be right about the "third location" that will emerge out of the conflict between these two deep stories, but they do see this as just another conflict. If the outcome they foresee comes to pass, its resolution will still demand further intervention and mediation that uses a Dialectical method of the sort that Lonergan developed. That is to say, what is needed is understanding, feelings, and deliberate commitments to the "deep story" of the emerging good, a story that contextualizes both American deep stories within a more comprehensive holistic comprehension of the good.

Hochschild seems to have overlooked the fact that there was nothing inevitable about the way that governmental policies to protect the environment and the climate were assimilated into the first deep story. Global warming and climate change themselves could have been cast in the place of governmental policies as the villains for the Cowboy to overcome. Indeed, Theodore Roosevelt himself saw dangers to the environment as a villain to conquer. Roosevelt as Rough Rider became one of the major incarnations of the Cowboy symbol, but he was nevertheless one of the strongest advocates for environmental preservation. Until the impact of John Sununu, both the Republican and the Democratic parties regarded environmental protection as compatible with the first deep story. America's deep stories were shaped by centuries of decisions calcified into cultural assumptions, but they could have been and still could be otherwise. The fight for positive change has to account for the influence of these deep stories on public thought, action, and division. In any society, therefore, addressing the cultural assumptions that shape political decisions has to be part of the work of environmental ethics. Lonergan's method of eight functional specialties was developed precisely to evaluate and criticize deep stories (originally, theological deep stories) and to provide Policies, Plans, and Implementations based upon these more refined assessments.

The Dialectic of Truncated and Holistic Notions of the Good

Lonergan's method of Dialectic is intended to identify fundamental sources of conflict as a basis for working toward effective resolution. Its primary mission is to "develop positions and reverse counterpositions."[62] In chapter 14, we saw that one basic form of conflict arises out of the differentiation between commonsense and scientific modes of knowing and evaluating, a differentiation that makes effective communication difficult. In this chapter we saw a different source: conflicting deep stories and the conflicting sets of feelings they cultivate.

I believe Hochschild is quite correct in using "feels-as-if" to characterize deep stories. They do shape and cultivate the feelings people have for values in powerfully formative ways. She argues, rightly in my judgment, that many political conflicts in the US are deeply affected by the conflicting feelings for conflicting values inculcated by the symbolisms of these deep stories.

Elsewhere I have argued that feelings are the first level in our consciousness of values, but feelings are not the final arbiter of objective valuations.[63] Feelings are not automatically objective assessments of values. Feelings initiate further processes of value reflection, deliberation, and value judgment. Objectivity regarding values comes about when and if these further processes of value reflection and deliberation are done thoroughly and carefully. While feelings are not automatically objective assessments of values, neither are they automatically irrational. While feelings can short-circuit the self-correcting processes of coming to judge and decide responsibly, judgments of value would be impossible without the feelings that initiate our reflective processes. Feelings inescapably play a role in the formation of value judgments, and they can contribute to objective assessment of values *if* they are supplemented with the demanding process of value reflection, the self-correcting cycle of questions and insights arising out of felt-values as it tends toward its proper limit.[64]

Lonergan's dialectical method is therefore relevant to the dialectical "shape shifting" of American political responses to climate change science, since the method's primary focus is on conflicts. In this case it is conflicting feelings about conflicting values and disvalues, as well as who and what are the true bearers of those values. Lonergan's method of Dialectic reveals the American deep stories to be *truncated*: they capture "positions" regarding values that are in need of further development, but whose development is held back by "counterpositions" regarding the good and the just. Nevertheless, each deep story is a partial contribution to the human quest for what is good and just.

While perhaps the differences between the deep stories are obvious, surprisingly both are in fundamental agreement in their commitment to at least one counterposition. *Both* deep stories cultivate feelings about the value of human life as oriented toward the American Dream. That dream is primarily cast in terms of material prosperity, esteem for success so defined, comfort and safety from interference, and dignity. Both nurture feelings for dignity in terms of respect due to one's individual efforts in forging one's individual way of life while achieving the dream.

These are not counterpositions for what they include so much as for what they exclude. What is missing are the symbolisms that cultivate feelings for the emerging good as a whole, not just for some parts of that good. There is an absence of symbols that evoke deeply felt valuation of communal cooperation in achieving a good life together as a continuation of the natural emerging good, as well as feelings for what and how large that community can be. What is missing are feelings for the good life defined not only in terms of the material goods that meet one's needs and wants but also defined in terms of a community forged not just individually but together, a community defined by excellences (*aretai*) of acting, feeling, and thinking directed both toward oneself and one's fellow creatures. What is missing is the cultivation of a "scale of value preferences" for the good as a *whole* in *all* its dimensions, including the natural environment. As such, neither American deep story is up to the challenges posed by climate change science.

Lonergan has provided something that goes beyond the limitations of the deep stories, and it is up to the challenge of guiding an adequate response to climate change. He did not provide a systematic, detailed program for meeting the challenges of climate change, something that is nigh impossible anyway. Rather than developing a systematic theory of the whole of the good in all its dimensions, Lonergan's approach was heuristic: he did not say what the solution to the challenges must be but instead provided structures to guide in the cultivation of feelings and the discovery and implementation of needed solutions. Lonergan's guidance comes in the form of his heuristic of the dynamic emerging good and especially the heuristic of the emerging "human good" embedded within it. This integral heuristic of the good has the advantage of anticipating the whole of the good in all its dimensions. As such, it includes both the genuine goods envisioned by the deep stories and the goods ideologically obscured by them.

It took centuries of dialectical modifications in symbolisms for the people of the United States to forge their deep stories about the American Dream into its current renditions. It will likewise take a long dialectical process to move from current deep stories to deep stories that symbolically communicate feelings for this larger, integral good that includes both human needs and the complex dynamism of climate, ecosystems, and all that they make possible. This larger vision of the good comes through what Lonergan calls "moral conversion." This is likely an unfortunate phrase, since people committed to both of the deep stories already understand themselves as seriously moral—and understand those in the opposing story to be less than fully moral. The difficulty is that these two conceptions of morality are truncated, not only because of what they cannot grant to each other but also because they cannot envision the whole of the good. What Lonergan meant by "moral conversion" is a conversion or reorientation of one's fundamental feelings and commitments from a limited range of goods to the totality of all goods in their complex and dynamic relationships with one another. Such holistic conversions do happen, as we have seen in the cases of Schweitzer, Leopold, Muir, and even Thomas Malone. People who become reoriented toward the good in its fullness leaven the dialectical process of addressing the climate and environmental challenges that lie ahead.

Studies of American Deep Stories

The following is a partial list of some important historical studies of the cultural deep stories (and particularly the symbol of the Cowboy) that have come to form the feelings of value preference for self-reliance and distrust of government among residents of the United States of America.

W. E. B. Du Bois, *The Souls of Black Folks: Essays and Sketches* (Chicago: A. C. McClurg, 1903)

W. E. B. Du Bois, *The Gift of Black Folk: The Negroes in the Making of America* (Boston, MA: Stratford, 1924)

Leo Marx, *The Machine in the Garden: Technology and the Pastoral Ideal in America* (New York: Oxford University Press, 1964)

Henry Nash Smith, *Virgin Land: The American West as Symbol and Myth* (New York: Vintage Books, 1970)

Garry Wills, *Nixon Agonistes: The Crisis of the Self-Made Man* (New York: New American Library, 1970)

Richard Slotkin, *Regeneration through Violence: The Mythology of the American Frontier, 1600–1860* (Norman: University of Oklahoma Press, 1973)

Michael Kammen, *People of Paradox: An Inquiry Concerning the Origins of American Civilization* (Ithaca, NY: Cornell University Press, 1980)

Richard Slotkin, *The Fatal Environment: The Myth of the Frontier in the Age of Industrialization, 1800–1890* (Norman: University of Oklahoma Press, 1985)

Richard Slotkin, *Gunfighter Nation: The Myth of the Frontier in Twentieth-Century America* (Norman: University of Oklahoma Press, 1998)

Andrew Delbanco, *The Real American Dream: A Meditation on Hope* (Cambridge, MA: Harvard University Press, 1999)

Heather Cox Richardson, *How the South Won the Civil War: Oligarchy, Democracy, and the Continuing Fight for the Soul of America* (New York: Oxford University Press, 2020)

Kristin Kobes du Mez, *Jesus and John Wayne: How White Evangelicals Corrupted a Faith and Fractured a Nation* (New York: Liveright, 2020)

Isabel Wilkerson, *Caste: The Origins of Our Discontents* (New York: Random House, 2020)

Chapter 16

Probability, Uncertainty, and Predictability in Science

According to Lonergan, history is a dialectical process in the sense that two or more basic but opposed principles keep modifying the flow of human events in conflicting ways. Actions done on the basis of one principle elicit modifications in behaviors and responses on the part of the other, followed by further counterresponses on the basis of the first, and so on. A "shape-shifting" pattern (as Wilkerson puts it)[1] of this dialectical process results as each principle attempts to maintain itself in response to the challenges posed from the other side. In chapter 14, I argued that the differentiation of scientific and commonsense patterns of knowing underlies certain aspects of the oscillations in public awareness about climate science and its ethical challenges. In chapter 15, I stated that there were two additional basic sources of the dialectic of climate denial: the deep story of antipathy toward governmental regulation in the United States, and a set of assumptions regarding certainty and uncertainty in science. In the previous chapter I examined the prevailing cultural deep stories that provide the felt-valuations that are driving that dialectic in American culture and politics, a dialectic that ultimately calls for a "moral conversion" to a holistic, integral notion of the emerging good. In this chapter I take up the other source of the dialectical conflict: the mistaken assumptions about the nature and certainty of scientific knowledge and the realities known by it.

Probability versus Uncertainty

We saw how John Sununu unleashed a broad network of climate denial by exploiting the uncertainty in the findings of climate science. Dunlap and McCright documented the rise of this denial network, uncovering the climate deniers' attack on "the scientific underpinnings of climate change policy" and highlighting "the crucial strategy of 'manufacturing uncertainty' they employ."[2] Although I am largely in agreement with the analysis of Dunlap and McCright, I would add one qualification: the denialists *exploited* rather than *manufactured* uncertainty in climate science.

The exploitation of the uncertainty in climate science relies upon widespread assumptions regarding certainty and uncertainty in science. It is widely assumed that all genuinely scientific statements are absolutely certain, but this is a false assumption. As it happens, uncertainty is not uncommon in modern science. Historian of science Theodore Rabb, for example, has written: "Scientists are never one hundred percent certain. . . . Even Newton has been proven to be wrong. That notion of total certainty is something too elusive ever to be sought."[3] In fact, certainty in modern science is a matter of degrees. As Lonergan put it, "Where the Aristotelian claimed certitude, the modern scientist disclaims anything more than probability."[4]

Scientific knowledge is seldom a matter of either absolute certainty or absolute uncertainty. Rather, scientific knowledge is a matter of greater or lesser probability. Scientific pronouncements can have minimal probability (barely more than an interesting hypothesis), or they can be somewhat likely, or they can be extremely probable, or they can be completely certain.[5] The climate deniers' strategy has been to collapse all degrees of probable scientific knowledge regarding the climate into the single category of "uncertainty," thereby casting extreme doubt on its worth for policy initiatives. They advance arguments that economic growth, prosperity, and jobs will be negatively impacted if policies to reduce GHG emissions are implemented, and that economic values should not be sacrificed on the basis of such uncertain knowledge. Of course, their unstated assumption is that their own forecasts about the impacts of GHG reduction policies upon the economy are absolutely certain. This is a false assumption, however, and their own forecasts about economic consequences are no more probable than the forecasts of climate change scientists.

This is where Lonergan's method of eight functional specialties can make important contributions. This attack on climate change science as uncertain rests upon what Lonergan called counterpositions regarding human knowing in general and scientific knowing in particular. These counterpositions comprise seriously mistaken assumptions about certainty, probability, scientific consensus, and trust in science. Counterpositions in general involve mistaken notions about reality, objectivity, and human knowing in the full and proper sense, and versions of these mistakes have played crucial roles in the dialectic of politics and climate change science.

Mistaken assumptions about what science is and what it can know have underpinned decisions to marginalize the findings of climate science. Because the scientific community had arrived at a consensus that its findings were highly *probable*, those findings were dismissed for being less than certain. Probability itself is dismissed as not being real. The persistence of these assumptions and the absence of a clearer understanding of the nature of uncertainty and probability in science was crucial to the rise of climate change denial during the George H. W. Bush administration. Resolution of this conflict requires what Lonergan called "intellectual conversion." Intellectual conversion entails, among other things, a proper understanding and conviction regarding of the roles of certainty, uncertainty, probability, predictability, and trust in modern science. That proper understanding and conviction must be based upon self-appropriation of the fullness of all the operations involved in scientific knowing. Intellectual conversion involves abandoning the mistaken but "exceedingly stubborn and misleading myth" about reality, objectivity, and human knowing in order to accept their full and proper senses, which are grounded in self-appropriation.[6]

Two Senses of Probability

The climate change deniers' strategy has been abetted by a common confusion about the meanings of uncertainty and probability. There are two distinct yet commonly intertwined meanings of "probability" in science that have long been working at cross-purposes throughout the history of modern science. Although Lonergan was not the only thinker to identify the confusion among these two meanings of probability (Karl Popper, for example, also discussed this issue at length, and Robert Boyle may have been the first), Lonergan nevertheless provided an acute way of clarifying the distinction between these two meanings.[7] Drawing upon his method of self-appropriation and the structure of human consciousness that it reveals, Lonergan distinguished between a certain kind of insight that grounds the first meaning of probability (i.e., "ideal frequency") on the one hand, and, on the other hand, a certain kind of judgment ("probable judgment") that lies at the root of the second kind of probability. Both kinds of probability play important roles in climate change science, and confusions between them have empowered climate change deniers.

Probability as Ideal Frequency

The first meaning of probability concerns the ideal frequency of occurrences and events, and there is a corresponding notion of uncertainty associated with this meaning. When scientists speak of average temperature or average rainfall or average number of hurricane-force storms in a year, they are invoking the meaning of probability as ideal frequency.[8] Lonergan argued that ideal frequency is the result of a particular kind of insight that begins with but goes beyond the empirical collection of statistical data on actual events and even beyond calculations based strictly upon this data alone.

The first stage in empirical statistical investigations is the formulation of a set of categories. In the case of climate change, relevant categories would be such things as the concentrations of GHGs in the atmosphere, temperatures, amounts of rainfall and snowfall, number of storms of varying intensities, size of ice sheets, and so forth. Once the categories have been agreed upon, the second stage in statistical investigations is to count the number of instances that actually fall into each of these categories in some region of the earth over some interval of time.

According to Lonergan, however, tabulation of the observations of actual instances in the categories is not the ultimate objective of statistical investigations. In the third stage, then, the actual *numbers* occurring in the categories are converted into actual *frequencies relative* to the total number of occurrences in a larger sequence or collection (called *actual relative frequencies*) in order to determine their relative representation in the whole. For example, the number of *actual dates* (say, 10) when the temperature was between 80° and 90° F (26.7°–32.2°C) in a particular region over the course of a year is converted to an *actual relative frequency* of 10/365. More generally, the number of actual events in a category divided by the number of possible events in a year would be represented by the generalized mathematical expression, N_f / N_p, first introduced by Laplace. Even this, though, is not yet a "probability" in the sense of an *ideal relative frequency*. In order to arrive at a probability in this specialized sense, the actual relative frequencies for temperatures have to be tabulated over the course of many years (or other appropriate time intervals). The sequence of these actual relative frequencies, Lonergan observes, will not necessarily follow any regular, "systematic" pattern over the course of time. Nevertheless, the use of proper statistical methods will enable scientists to have insights into hypotheses about *ideal* relative frequencies from which the yearly *actual* relative frequencies deviate only randomly.[9] In other words, statistical methods enable scientists to arrive at insights that abstract ideal relative frequencies (probabilities) from the random messiness of actual relative frequencies. For example, after Gregor Mendel tabulated his actual frequencies of certain characteristics in peas (as 2.98/4) he went on to declare that the probability in the sense of ideal frequency was 3/4, and he proceeded to postulate dominant and recessive hereditary characteristics to explain this exact number.[10] Initially, however, insights into ideal relative frequencies (probabilities) are only hypotheses in need of proper verification.

Probability as ideal relative frequency is important for climate change science. Scientists calculate the probabilities of rainfall, powerful storms, heat waves, droughts, wild fires, warming of oceanic waters, melting of ice sheets and permafrost, and so on. All these play a role in scientific forecasting of the future changes to the climate over the coming years. Climate is basically what Lonergan calls a "schedule" or list of probabilities of different kinds of events.[11] That is to say, climate is characterized not just by the ideal relative frequency of temperature between 80° and 90° F (26.7°–32.2°C) in a region. It is the set of all the ideal relative frequencies associated with the whole set of temperature intervals for that region, along with similar sets for other climate phenomena.

When climate is unchanging, the list of ideal frequencies and averages is unchanging, even though the actual frequencies do change from year to year. The actual occurrence of events is weather, and weather changes from day to day and year to year. Unlike weather, climate does not consist of actual events but is instead the schedule of ideal frequencies (or probabilities) of these events. Therefore, the climate can be unchanging although the weather is constantly changing nonsystematically. As such, an actual amount of rainfall or temperature might rise above the ideal frequency in one or a few years, but it will eventually (though nonsystematically) be offset over the course of the next few years by some actual declines. Climate change, on the other hand, is a dramatically different kind of change. Climate change means that the list of ideal frequencies itself is changing systematically. This is the very claim that scientists are now advancing.

Probability as a Matter of Judgment

As with any scientific hypothesis, the hypothesis that climate probabilities and averages are changing was initially no more than that: a hypothetical possibility. This is borne out in the history of the early stages of climate change science where hypotheses about changes in average temperatures were under investigation. As with any set of insights formulated into hypotheses, the hypothesis of climate change called forth the further question "But, is this so?" Questions of this sort bring about a transition from the operations proper to what Lonergan calls the level of intelligence—in which one seeks insights—to the additional operations at the higher level of reflection and judgments of fact. Such questions bring about a transition from using the methods of statistics to form plausible hypotheses about ideal relative frequencies to using other methods to determine whether those hypotheses are correct or at least probably correct.

The importance of questions and the self-correcting process in science provides the basis for clarifying the second meaning of probability: judgment of a hypothesis as probably correct. The empiricist counterposition on knowing exerts a subtle but subversive influence here, for empiricism assumes that scientific knowing is principally a matter of gathering sense data.[12] The empiricist counterposition has led to the mistaken notion that an increase in the amount of sense data in support of a hypothesis will lead to an increase in the probable truth of the hypothesis. This has led to a confusion of probable judgments with the first notion of probability as ideal frequency. This confusion traces back to Laplace's definition of the first meaning of probability. If probability is the numerical fraction of favored events over all possible events ($P = N_f / N_p$), it would seem that increasing the amount of empirical data (N_f) in favor of the hypothesis would increase the probability of the hypothesis. This assumption that the probability of a hypothesis increases with additional data in its support has come under considerable criticism by Popper, among others, but the counterposition has proved highly impervious to such criticisms.

According to Lonergan, sense data do *not* give human beings immediate contact with reality, nor do they determine the probability of a judgment. As he put it, the relationship of human knowledge to reality is not given immediately through sensations. Rather, reality is only "mediate in the data of sense." He explains that "while it is true enough that data of sense result in us from the action of external objects, it is not true that we know this by sense alone." Sense data provide only a "mediate" or indirect, partial contribution to human knowledge of empirical reality. Only as the human inquiring spirit "makes use of data in promoting cognition to knowledge" of reality does empirical data contribute to the advance from mere hypothesis toward correct judgments.[13]

In other words, "Is it so?" questions give rise to the self-correcting processes of reflection, which begin to discern what conditions would have to be fulfilled in order to make a reasonable judgment about the correctness of a hypothetical insight.[14] Reflective process will work out which sense data would count as relevant conditions and which data should be ignored. But no finite number of data points can ever verify a hypothesis as the one and only correct understanding. Therefore, processes of reflection also discern which conditions beyond empirical data, such as additional insights, need to be fulfilled before a definitive judgment can be reasonably pronounced. These processes of reflection, not sensation alone, determine the relevance of some sets of sense data and the irrelevance of other sense data. The self-correcting reflective processes guided by the "Is it so?" questions—not sensations—are what is most basic in knowledge of reality. Reality is known not through sensations alone but through correct judgments that incorporate sensations and insights into a "virtually unconditioned" synthesis that serves as the reasonable ground for a judgment of fact.[15] Since sensations alone do not constitute correct knowledge, neither does the mere addition of more sensations alone make judgments more probable.

The processes of reflection advance from mere hypothesis toward verified judgments of fact by acquiring answers to these further questions. The self-correcting cycle of knowing consists of the sequence of data, questions, insights that uncover further data leading to further questions, and further insights that complement and correct earlier ones. New insights help investigators notice empirical details they had previously overlooked. New insights lead to the construction of instruments that produce new and otherwise inaccessible data. As this cycle repeats, the growing accumulation of insights can tend toward a limit where there would be "no further pertinent questions."[16] Once such a limit is reached, the accumulated insights, some of which have corrected earlier ones, constitute a correct understanding of the data. The fulfillment of all of these requisite conditions leads to a judgment that is both correct and completely certain.[17]

As Rabb has pointed out, however, attaining absolutely certain answers to all questions in the realm of science is rare. Probable judgments occur when one knows that some but not all of the requisite conditions have been fulfilled. This may include awareness that some sense data have not yet been gathered, but it almost always includes recognition that there is also need for further insights to answer some of the as yet unanswered questions. *The relative probability of two competing hypotheses, therefore, is measured more by the range of questions each has answered than by the quantity of sense data gathered in support of each.* One knows this by self-appropriation, which pays attention to the subtle stirrings of questions that remain unanswered.

For Lonergan, therefore, a probable judgment is a judgment that one hypothesis or set of insights is more probably correct than others. Probable judgments are comparative judgments. As insights accumulate answering ever more questions, the later insights correct the deficiencies of the former and expand the probability of the judgments based upon them. Probable judgments situate hypotheses and the insights that they formulate into a serial relationship with one another, with later hypotheses being more probable because they have answered questions left unanswered by the former.

This account of probable judgments in general is also true in particular of probable judgments about the first type of probabilities. An insight that p/n is the correct ideal frequency (probability) for some property P in some population of n events will be initially just a hypothesis. As further investigations are undertaken, however, and as questions about alternative fractions that could be the ideal frequency are scrutinized and dismissed, p/n is gradually judged to be not merely hypothetical but ever more probably true. When an insight into some alternative ideal relative frequency, p'/n, more successfully meets the further questions, it is judged to be more probable. It becomes a "probable probability" because of its superiority in meeting the further questions to which the hypothesis *p/n* gave rise. That superiority consists in the fact that it has answered questions raised but not successfully answered by its predecessors.

Lonergan therefore analyzed the series of successively more probable judgments of fact in terms of the self-correcting dynamic of questions and insights: "[The] self-correcting process of learning is a circuit in which insights reveal their shortcomings by putting forth deeds or words or thoughts, and through that revelation prompt the further questions that lead to complementary insights."[18] The self-correcting process is motivated and guided throughout every cycle by "the very spirit of inquiry that constitutes the scientific attitude," as Lonergan puts it.[19]

Of course, this self-correcting process also involves the gathering of additional empirical data; this is quite obvious to scientists and nonscientists alike. Less obvious, however, is the equally essential self-correcting process of further questions followed by insights that modify earlier ones. This account of the self-correcting process provides the basis for Lonergan's analysis of probable judgments. As he put it:

> The probable judgment results from rational procedures. Though it rests on incomplete knowledge, still there has to be some approximation towards completeness . . . the self-correcting process of learning consists in a sequence of questions, insights, further questions, and further insights that moves towards a limit in which no further pertinent questions arise. When we are well beyond that limit, judgments are obviously certain. When we are well short of that limit, judgments are at best probable . . . because the self-correcting process of learning is an approach to a limit of no further pertinent questions, there are probable judgments that are probably true in the sense that they approximate to a truth that as yet is not known.[20]

Noteworthy in Lonergan's remarks is his insistence that even though probable judgments fall short of certainty, they are nevertheless the results of rational procedures. The rational procedures of the self-correcting process may include but are not confined to the strict rules of logic. Logic guarantees that rigorously deduced conclusions will be true if premises are true. But logic alone cannot be the method for arriving at true ultimate premises. For that the self-correcting process is needed, and in science those premises and consequent conclusions will be increasingly probable but seldom absolutely certain.

The predictions by the climate science community were probable judgments in the second sense of probability explained in this section. They came out of the hard work put in by thousands of scientists as they found answers to countless further pertinent questions over the course of a century. As Weart astutely said of the 1977 NAS report: "The few pages of text and graphs were the visible tip of a prodigious, unseen volume of work. Many thousands of people in many countries had spent much of their lives measuring the weather, while thousands more had devoted themselves to organizing and administering the programs, improving the instruments, standardizing the data, and maintaining the records in archives."[21] The same could justifiably be said for all those individuals involved in utilizing theories to understand the connections among the data, as well as those who designed, coded, ran, and refined the computer models of climate.

Relevance to Climate Change Science and Denial

What is the source of the high degree of probability that climate scientists now invest in their predictions about the changing climate probabilities? Weart remarks that "by following how scientists in the past fought their way through the uncertainties of climate change, we can judge why they speak as they do today."[22] The history of climate science is the history of the growth in probable judgments about the complex realities of earthly climate. The relevance of the historical summaries in the preceding chapters now becomes apparent. Weart narrates the all-important process through which scientists "had begun to question the comfortable old belief that climate was regulated in a stable natural balance, immune to human interventions."[23] He and the other historians carefully traced the complex sets of questions arising out of new observations that led to ever more refined, sophisticated, and complex insights into climate. Careful observation of "erratic" rocks led to questions about how they came to be at their odd locations. Insights into the possibility of glacial forces and movements led to provisional explanations that were gradually expanded and refined over decades through further questions, insights, and observations. The insights about glacial advance and recession gave rise to theories about a series of ice ages and then to questions about their causes. These led to Arrhenius' insights into the possibility that variations in water vapor and CO_2 could be the result of atmospheric heating and cooling and possibly the cause of ice ages. Arrhenius' work raised further questions for Callendar, which in turn led to his insights and experiments that made the influence of CO_2 ever more probable. Among other things, Callendar's questions about the adequacy of CO_2 absorption to account for heating led to questions and insights about absorption at longer wavelengths. As a result of this series of further questions and insights, his judgments about the role of CO_2 in warming of the atmosphere became far more probable than those of Arrhenius. As Fleming remarks, "Scientists, inspired by Callendar, began to investigate in greater detail the linkages between rising CO_2 levels and rising temperature."[24] These and other complex feedback phenomena in climate posed formidable challenges for scientific investigation. Between 1950 and 1977, scores of computer scientists labored to develop computerized climate models to incorporate such complexities. Initially the computers and the software were woefully inadequate, leading to questions and insights that could incorporate more complex factors. Over the course of a quarter century computer scientists immersed themselves in the self-correcting cycle of questions, insights, further questions, and further insights, gradually improving the accuracy of their models. Among other things, they had to convince other scientists that what they were doing was legitimate empirical science.

Even though the scientific community knew in 1977 that there still remained further questions, a consensus had emerged that the hypothesis regarding the human causes of climate change probabilities was probably correct. Further questions, observations, and insights over the following decades led to judgments that were even more probable. As Weart puts it: "It was not the weight of any single piece of evidence that convinced them, but the accumulation of evidence from different, independent fields . . . increased haze and jet contrail clouds, ancient catastrophic droughts, fluctuating layers in ice and in seabed clays, computer calculations of planetary orbits and of energy budgets and of ice sheet collapse: each told a story of climate systems prone to terrible lurches. Each story, bizarre in itself, was made plausible by the others."[25] In Weart's statement, "accumulation of evidence" may seem to suggest large quantities of bare, uninterpreted sense data. As the historical research by him and others show, though, "evidence" really means data *as* integrated into an ongoing process of ever more insights, responding to further questions and ever more probable judgments generated by the self-correcting cycle. As that cycle adds ever more improved insights, they become "evidence," so that as the cycle progresses, it turns bare data into ever more "evidence." As Weart observes, "Scientists therefore rarely accept a result until it is confirmed by wholly different means."[26] Historians of climate change science have critically traced over a century of self-correcting cycles of observations, questions, and insights that have produced ever more probable judgments. These judgments were not absolutely certain, but neither were they merely uncertain. Their increasingly higher degrees of probability and rationality were won through the countless number of questions they answered, eventually giving them so high a status of probability that they and their ethical implications deserved serious attention.

By way of contrast, the network of climate deniers mischaracterizes the series of increasingly more probable judgments as merely uncertain. At the same time, they rely on a small number of maverick scientists who offer a few sporadic, isolated findings and interpretations that seem to be at variance with the prevailing scientific consensus. While great advances in science have often come from just such nonconformists, a great many of the questions frequently raised by the climate change deniers have already been considered and answered at previous moments in the history of climate science—hence Weart's emphasis on the importance of knowing the history of science.

In addition, the deniers also assert that our economic prosperity and employment levels will suffer devasting consequences, and they make such assertions as though these were absolutely certain claims. That there will be economic consequences for shifts in human economic and social patterns is undeniable. That questions regarding these economic consequences must be answered fully is ethically imperative. But it is hypocritical to reject climate science as failing to provide absolutely certain, precise predictions while pretending that their own social and economic predictions have such precision. Lonergan's analysis of probable judgments helps to expose this hypocrisy.

Probability and Prediction

There is one other counterposition that has reinforced denial of climate change: the confusion of probability with uncertainty about predictions.

This confusion was introduced into modern thought by the Marquis Simone Pierre de Laplace in his *Philosophical Essay Concerning Probabilities*. As he put it:

> We may regard the present state of the universe as the effect of its past and the cause of its future. An intellect which at a certain moment would know all forces that set nature in motion, and all positions of all items of which nature is composed, if this intellect were also vast enough to submit these data to analysis, it would embrace in a single formula the movements of the greatest bodies of the universe and those of the tiniest atom; for such an intellect nothing would be *uncertain* and the future just like the past would be present before its eyes.[27]

In this way, Laplace introduced the misleading equation of certainty with exact predictability.

Laplace acknowledged that, at present, scientists are ignorant of the exact laws governing the motions of objects in space and of the precise measurements of the "given circumstances." Since it was "impossible for us to announce [future] occurrence with certainty," scientists had to be content to work with probabilities.[28] In other words, the whole point of probability in science is to partially offset the temporary uncertainty in predicting future events. Laplace believed real knowledge would eventually replace those probabilities, which he regarded as mere masks for uncertainties.

As we have seen in the previous sections, this is a confusion on the part of Laplace of the two radically distinct notions of probability. More than anyone else, Laplace is the person responsible for propagating this confusion. In Lonergan's analysis, *if* a numerical probability is judged to be certainly true, then its ideal relative frequency is a constitutive component of reality. The set of probabilities associated with the climate would truly characterize the reality of climate. If one judged that that set of probabilities were certainly changing, this would be a true judgment about the reality of climate change. Probabilities as ideal frequencies are not temporary placeholders for the day when absolute certainties are attained. Probabilities as ideal frequencies, when completely verified, are genuine contributions to our knowledge of the real world.

Of course, climate scientists do not claim that their judgments about the ideal relative frequencies related to climate change are absolutely certain. They know those probabilities to be probably true. They know the current estimates to be much more probable than even the numbers available in 1977. Still, it would not make sense to ask for a *number* expressing the probable truth of the probabilities accepted by the consensus of climate scientists either in 1977 or today. That would be a confusing of the two different meanings of "probability."

Climate deniers have exploited this supposedly problematic lack of absolute certainty, by mistakenly associating certainty with precise predictions of future events. The lack of precise prediction is especially evident when the events are as numerous, as complex, and as intertwined as are the dynamic systems of climate. This, of course, is something that climate scientists freely admit: they do not pretend to predict with precision the exact paths of hurricanes, the precise locations and extents of wild fires, or precisely which species will become extinct and exactly when, or the exact date and time when the sea level of Boston Harbor will rise to one foot above its current level. What they predict is changes in probabilities and averages of events, from which actual events will deviate only nonsystematically. They predict not precise, specific events but probabilities, and they do so not with certainty but with a fairly high degree of probability. As Lonergan put it, "If probabilities must be verified, it also is true that there is a probability of verifications" of these probabilities.[29]

Even if scientists knew probabilities (ideal frequencies) with absolute certainty, precise predictions with certainty would still remain an elusive if not impossible goal. This is due to the phenomenon of dispersion, which can be illustrated by the way that meteorologists report their predictions of the tracks of hurricanes. Meteorologists using different computer models typically predict different paths, intensifications, potential damage, and so on. Usually each of these paths is associated with a different degree of probability. Yet all of these predictions are based upon the same sets of data.

This dispersion of probable hurricane paths is due to the fact that the knowledge of the conditions that affect hurricane progression is itself a matter of probabilities. The measurements of temperature, air pressure, water saturation, wind, and water currents for specific regions in the ocean are situated within a limited range, each accompanied by its own probability. These probabilities are ideal frequencies from which precise actual conditions diverge nonsystematically. Hence computer models predicting storm formation and development will diverge from one another depending on which nonsystematic divergence of conditions is used in the model. In addition, because of the nature of complex dynamic systems, the longer the time span involved, the more the trajectories and the intensities of the possible storms will diverge from one another. These divergences will become pronounced, for as the computer modelers discovered, the complex dynamic systems that constitute climate patterns are highly sensitive to minute deviations in initial conditions (see chapter 13). As Lonergan says, a "nonsystematic process . . . leaves internal factors all the freer to interfere with one another."[30] What is true for the impossibility of precisely predicting the course of hurricanes is all the more true of global climate changes in general.

Laplace and many others would no doubt respond that dispersion and unpredictability are due to the fact that probabilities are mere placeholders. They would claim that ideal relative frequencies are used because of the lack of precision in measuring the initial conditions that affect hurricanes and other climate patterns. They allege that once precise measurement of conditions replaces mere probabilities, exact predictions will become possible. Unfortunately for them, complex dynamic systems and the overall nonsystematic character of global climate make precise prediction both practically and theoretically impossible.[31] Hence the expectation that real science will deliver precise, absolutely certain predictions is a deception based upon counterpositions regarding scientific knowing.

Responding Ethically to Probable Probabilities

The forgoing discussion of probability and predictability might seem to leave us in a real quandary for deciding how to respond ethically to the consensus statements of climate scientists. Even with the preceding clarification about the two meanings of probabilities, if climate science delivers no more than probable probabilities, and if probabilities do not deliver exact predictions, what possible relevance could this science have for ethical deliberation and action?

The answer has to do with the nature of probabilities themselves. Probabilities do not regard individual events or processes. Probabilities regard collections of events. It is not meaningful to speak of the probability of a single event. Laplace's definition $P = N_f / N_p$ explicitly refers to a collection of N_p members. If $N_p = 1$, however, the definition becomes fairly meaningless. Moreover, if it is certainly or even probably true that the fraction P *is* the probability of events of type f occurring, then they must occur eventually. Otherwise P would not be an empirically meaningful number. All this implies that even if events of type f do not occur at specified places and times, they will eventually occur numerous times over a sufficiently long span of space and time.

This of course is exactly why climate scientists have raised ethical alarm. What they forecast is not where or when climate disasters will occur. Rather, they predict that the frequencies of disastrous occurrences are increasing at an alarming rate as time moves forward. This means that increasing numbers of natural ecosystems and human communities will suffer as a result, even though the exact times and places of these disasters are not predictable. An ethics that is concerned only with individual outcomes may not be impressed by our analysis, but a holistic ethics will be. Knowledge that the ideal relative frequencies of increasing devastation to natural ecosystems and human communities is highly probable would lead holistic ethical thinkers and actors to creative insights followed by ethical feelings, deliberations, judgments of value, decisions, and actions to change those ideal relative frequencies for the sake of the whole, dynamic planetary environment and the human communities situated within it.

Even an individualistically oriented ethic is not wholly indifferent to this kind of scientific knowledge about probabilities, though. As Weart observes: "If there is even a small risk [probability] that your house will burn down, you will take care to install smoke alarms and buy insurance. We can scarcely do less for the well-being of our society and the planet's ecosystems."[32] Ordinary people in business and family life make decisions based on probable probabilities, that is, the best available understanding of probabilities as ideal frequencies. Ordinary people do not know for certain what will happen. They arrive at commonsense estimates and judgments on the basis of probabilities, knowing full well that they themselves might never receive an insurance payout. That does not stop them from making comparative judgments of value such that it would be better to risk putting a limited amount of money into insurance that might never pay off than to suffer a devastating loss that they could not afford to their health, home, or transportation.

This is the type of ethical decision confronting the world. Perhaps we could wait for a few more decades to get even more probable judgments that answer more of the currently unanswered questions. But as James Hansen has pointed out, if the current probable judgments are correct, the results of waiting for more certain judgments will have catastrophic consequences. Weart adds:

> Of course, climate science is full of uncertainties, and nobody claims to know exactly what the climate will do. [But] we can conclude (with the IPCC) that it is *very likely* that serious global warming, caused by human actions, is coming in our own lifetimes. The probability of harm, widespread and grave, is far higher than the probability of many other dangers the people normally prepare for. The few who contest these facts are either ignorant or so committed to their viewpoint that they will seize on any excuse to deny the risk.[33]

He further notes that when the 1990 IPCC report "said the future was uncertain, that did not mean the risk to climate should be ignored. The experts had pointed out economically sound ways to get a start on reducing the risk, a broad hint that governments should begin to act."[34]

Climate change deniers have largely been motivated by bargaining that the short-term economic costs are not worth the long-term probable consequences. As we saw in chapter 15, Secretary of State Baker held that we cannot afford to wait until all of the uncertainties about global climate change have been resolved, at least until he was silenced by Sununu. It was a position that acknowledged some degree of uncertainty (probability of judgment) in science but nevertheless addressed those probabilities in the ways people of common sense do when they make decisions to buy insurance policies. This is exactly the right ethical position.

For an unchanging reality and an unchanging, absolutely certain knowledge base, a static ethical system may suffice, but a holistic and heuristic framework of the emerging good is needed to guide changing ethical knowledge and action that has to respond to a dynamically changing reality (climate) and a continually evolving advance in the probable scientific understanding of probabilities (ideal frequencies). In the next chapter, we turn to the phenomenon of scientific consensus and the role of trust and belief in science. In the final two chapters, we examine recent developments in the formulation of other holistic, heuristic frameworks by Pope Francis and the United Nations in comparison to that of Bernard Lonergan.

Chapter 17

Scientific Consensus, Trust, and Belief

Toward the end of his history of climate change science, Spencer Weart poses the question "When should we *believe* that scientists are giving us reliable information about the world?"[1] This question is at the heart of the crisis of denial and the ensuing dialectical history of climate change science, ethics, and politics in both the US and the world at large. Even if deniers and others could be persuaded to give up rigid insistence on absolute certainty and take guidance from the best available scientific understanding (i.e., the probable judgments of the scientific community), the movement of climate change denial would still have a powerful basis: *mistrust*. The highly probable judgments about global warming and climate change are the consensus of the scientific community, so, at bottom, the crisis of climate change denial is a failure of trust and belief in the scientific community itself.

Most scientists and many people in general might object that belief has no place in science at all. In debates and discussions, scientists and the activists who follow their findings both insist that their claims are based upon facts, not mere beliefs. They accuse those who deny the realities science has discovered and verified at least probably as being guilty of bringing belief into matters where it has no place. This, however, is a weak response to the deniers, for it ignores the pervasive roles that belief and trust actually do play in science. Such responses fail to go to the root of the crisis and to make convincing impacts upon the mindsets of deniers. What is needed is the more discerning analysis of knowledge and belief that comes out of Lonergan's method of self-appropriation.

Trust, Belief, and Scientific Consensus

Although activists who appeal to the findings of science adamantly insist that their exhortations are based upon facts rather than beliefs, this is only partly true. Climate change and environmental activists themselves trust scientists; they therefore believe the facts upon which they base their activist arguments. Few activists have used scientific instruments to make observations, have performed experiments, or have used equations or written computer algorithms to calculate the results that they cite in support of their calls for action.

Trust and belief also play major roles in the practices of scientists themselves. At the earliest stages of their education, scientists trust their teachers and textbooks and lab manuals. Women and men begin their initiations into scientific disciplines by believing that what they are being taught is true. Trust and belief at this early stage gives them the space to do the hard work of acquiring the thousands of insights they will need to understand the ideas that have been bequeathed to them by previous scientists so that they can pass exams and prepare to engage in their own research. Scientists also trust the suppliers of lab equipment, reagents, and specimens. Although they do test some samples of the items supplied, they do not and cannot first test each and every item; otherwise, they would have no untainted items for use in the experiments themselves.

Scientists also trust one another and believe the results reported by other scientists. While it is true that some scientist somewhere does herself or himself have direct, "immanently generated knowledge"[2] of some particular scientific fact, no one has direct knowledge of all the facts that scientists have discovered about climate change or any other scientific specialty over the course of more than a century. They trust one another to be honest in reporting what they know and what they do not know. Even more do they trust the institutionalized vetting processes (e.g., refereed journals and conferences and what Weart calls "the flexible, freewheeling groups that scientists had perfected over centuries"[3]) that keep scientists honest and expose deviations from proper scientific procedures.

Lonergan has observed, "Science is often contrasted with belief, but the fact of the matter is that belief plays as large a role in science as in most other areas of human activity."[4] His father was a land surveyor of the Canadian wilderness, so Lonergan used map surveying to illustrate his point. He wrote: "There is a narrow strip of space-time that each of us occupies," and we rely on maps to know about the regions we have not yet visited. Maps are based upon the work of surveyors, in whom we place our trust, but no surveyor surveys the entirety of a province or nation, so the "accuracy of the whole [map] is a matter not of knowledge but of belief, of the surveyors believing one another and the rest of us believing the surveyors."[5] Likewise no scientist has personal access to more than a narrow strip of data or has performed more than a miniscule portion of the calculations that make up the totality of what is known in any field of science. Lonergan points out that scientists do not "spend their lives repeating one another's work."[6] They can only make their own contributions by believing the reported results of others to be true (or at least very probably true). They believe the results of others and build upon them by adding their own further insights and observations, thereby increasing the probability of scientific judgments.

Weart describes the intricate process of building scientific trust eloquently:

> For the process to work, scientists must trust their colleagues. How is this trust maintained? Telling the truth is important, but it is not enough: although scientists rarely cheat one another, they easily fool themselves. The essential kind of trust comes from sharing a goal, namely the discovery of reliable knowledge, and from sharing principles about how to pursue that goal. One necessary principle is tolerance of dissent, encouraging every rational argument to be heard in public discussion. A second principle is *limited consensus* working to agreement on important points, even while disagreeing on others.[7]

The reliability of the probable judgments of climate science is the achievement of this kind of consensus. It is a consensus comprising what Lonergan calls a "symbiotic fusion" of beliefs and immanently generated knowledge.[8] In this interdependent fusion, the vast majority of the community trusts and believes the immanently generated results produced by some other scientist or research group.

Still, immanently generated knowledge is one thing and belief another. By immanently generated knowledge Lonergan does not mean merely seeing for oneself. Rather, immanently generated knowledge consists of the certain or at least probable judgments one arrives at oneself through the intricacies of the self-correcting cycle of knowing in which one's own judgments depend exclusively upon one's own insights answering one's own questions about only the sense experiences that one has available to oneself. No one scientist, however, has arrived at all of the judgments about climate change by her or his own immanently generated knowledge alone.

> Who made the discovery of global warming—that is to say, the discovery that human activities are making the world warmer? No one person, but a sprawl of scientific communities. Their achievement was not just to accumulate data and perform calculations, but also to *link* these together. This process was so complex and so important that its late stages were visibly institutionalized: the workshops, reviews, and negotiating sessions of the [IPCC]. The discovery of global warming was patently a social product, a *limited consensus* of judgments arising in countless discussions among thousands of experts.[9]

In order for any scientist to affirm the great many substantiated judgments about climate change, she or he has had to fulfill the requisite conditions for those judgments by believing a great many results put forward by thousands of others.

This is what philosophers and sociologists mean when they speak of science as a "social construction." Studies of science as social construction point out that no one scientist has personally experienced all the relevant empirical data for her- or himself, had all the insights that answer questions for intelligence about that data, nor formed all the judgments about the correctness of those insights. In addition, social constructivists argue, all data is already interpreted according to some socially agreed upon norms. Their implicit if not explicit conclusion, therefore, is that decisions, not empirical data, are the only discernable grounds for the judgments held by groups of scientists. It is further assumed that these are more or less arbitrary and unexamined decisions, motivated perhaps by the social pressures of the scientific community.[10]

In response, I join Weart in insisting against this conclusion that "we should not call [a scientific consensus] *nothing* but a social product."[11] It is not true that beliefs rest only upon arbitrary decisions lacking rational grounds. Parallel to the self-correcting cycle of immanently generated knowledge, Lonergan identified a self-correcting cycle of critical believing. This cycle consists of finding genuine answers to questions about

> the trustworthiness of a witness, a source, a report, the competence of an expert, the soundness of judgment of a teacher, a counselor, a leader, a statesman, an authority. The point at issue in each case is whether one's source was critical of his sources, whether he reached cognitional self-transcendence in his judgments of fact and moral self-transcendence in his judgments of value, whether he was truthful and accurate in his statements. Commonly such questions cannot be answered by direct methods, and recourse must be had to indirect. Thus, there may be more than one source, expert, authority; they may be independent and yet concur. Again, the source, expert, authority may speak on several occasions: his or her statements may be inherently probable, consistent with one another and with all one knows from other sources, experts, authorities. . . . Finally, when everything favors belief except the intrinsic probability of the statement to be believed, one can ask oneself whether the fault is not in oneself, whether it is not the limitation of one's own horizon that prevents one from grasping the intrinsic probability of the statement in question.[12]

This process is eminently reasonable because it engages ever further questions that are pertinent to whether one should trust person *P* who is claiming that *S* is factually or at least probably true. Taking seriously the retinue of questions listed by Lonergan and withholding judgment until these and other relevant questions are answered satisfactorily together constitute reasonable, critical believing. But notice, the set of further questions is not about statement *S* itself. Addressing all the pertinent questions about *S* is how *P* presumably reached her or his immanently generated knowledge that *S* is so (or at least probably so) in the first place. Rather, the critical believer has to ask all the pertinent questions about *P*, about whether *P* is to be trusted to have done the research carefully and objectively, and on this basis, whether *S* should be believed.

Finally, note that among the reasons Lonergan gives for trusting others about their claims is the self-correcting process of knowing itself.

Even dim awareness of how it operates within oneself enables one to recognize that it can and frequently does operate in others as well. Lonergan observes that "other inquirers may have frequently appealed to the same source, expert, authority, and have concluded to the trustworthiness of the source, the competence of the expert, the sound judgment of the authority."[13] The self-correcting process of inquiry will eventually turn up oversights. It will also gradually support probable judgments by others that have survived the self-correcting process. And it may even eventually turn up reasons why *P* might not be trustworthy, at least in the field of knowledge where the claim *S* is being made.

Independently of Lonergan, Weart describes at length how critical trust and belief have occurred in the history of climate science. "Long gone were the days when the great questions of climate could be profitably studied by a few people. . . . The job now was to find precise answers to *countless specific questions*."[14] He continues:

> Maintaining *trust* is more difficult where the social structure is fragmented. A community in one specialty cannot thoroughly check the work of researchers in another, but must find a few people who *seem reliable* [which is a probable judgment of value] and accept their word for how to view the latest results. The study of climate change is an extreme example. Researchers cannot separate meteorology from solar physics, pollution chemistry, and so forth. The range of journals they cite in their footnotes is remarkably broad. So many different factors influence climate! But this complexity imposes difficulties on those who try to reach solid conclusions about climate change. Establishing a level of *reliability* depends on a *process of checking and correction* by dozens of scientific communities, each dealing with its own piece of the problem.[15]

Independently of Weart Lonergan wrote:

> New results, if not disputed, tend to be assumed in further work. If the further work prospers, they begin to be regarded with confidence. If the further work runs into difficulties, [the earlier results] will come under suspicion, be submitted to scrutiny, tested at this or that apparently weak point.

> Moreover, this indirect process of verification and falsification is far more important than the initial direct process. For the indirect process is continuous and cumulative. It regards the hypothesis in all its suppositions and consequences. It recurs every time any of these is presupposed. It constitutes an ever increasing body of evidence that the hypothesis is satisfactory. And, like the evidence for the accuracy of maps, it is operative only slightly as immanently generated knowledge but overwhelmingly as belief.[16]

Weart describes how scientific consensus of trust and belief is generated by diverse and even opposed teams, methods, and sides of debates. "You cannot point to a single observation or model that convinced everyone about anything. . . . It is more like a crowd of people scurrying about, comparing notes, shouting criticisms across the hubbub."[17] This description aptly conveys the "limited consensus" of the scientific community that affirms the highly probable judgments about global warming and climate change. Weart remarks that this limited scientific consensus was achieved by the late 1980s (although several scientists individually arrived at the same judgments a decade earlier).

Yet this is a *limited* rather than unanimous consensus. It is limited to a substantial but not universal group of scientists who accept the probable judgments. Weart observes that "tolerance of dissent, encouraging every rational argument to be heard in public discussion," is a defining feature of a limited scientific consensus.[18] This is because a significant number of the scientific judgments about climate change are probable, not absolutely certain. This means that members of the consensus are aware of the further questions that remain to be answered, and they respect scientists who withhold their assent until those questions are answered. "Future climate change in this regard is like electrons, galaxies, and many other things not immediately accessible to our senses. All these concepts emerged from a vigorous struggle of ideas, until *most* people were persuaded to say the concepts represented something real."[19]

Finally, the consensus is also limited by the sheer complexity of the climate itself. There are so many complex, interacting factors in climate, and the exact paths of climate processes are highly sensitive to minute differences in physical conditions. This means that there always remain further questions to be answered, so the judgments of the community will be probable for the foreseeable future. That said, they *are* probable, *not merely* uncertain, and the probability of the judgments in the consensus has been gradually increasing.

At bottom, then, scientists trust one another because they trust the scientific methods that they have acquired in their education as scientists. They believe the reports of those whom they trust to have followed those methods assiduously. Their most fundamental trust is in the methodical, self-correcting cycle that repeatedly generates ever more sense data, further questions, further insights, and ever more probable judgments of fact. As they advance from novice to expert scientists, they come to understand for themselves the reliability of the methods and findings they originally believed as students. Such realizations ground value judgments that they were right in originally trusting their teachers and books—until further questions arise that call those into question themselves. While most scientists trust the methods that have been developed by centuries of predecessors, Lonergan provided a basis for both knowing and refining judgments about the trustworthiness of the work and teaching of others through his own method of self-appropriation.

Alienation of Trust in Science and Deep Stories

Scientists as well as activists tend to underestimate the pervasive importance of trust and belief in their own practices. They therefore tend to underestimate the importance of building trust with people of common sense so that they will believe what scientists have discovered to be probably true about climate and the natural world in general. The alienation of trust between scientists and large portions of the public is at the heart of the climate change crisis. This alienation of trust has more recently extended beyond climate change to other scientific claims about viruses and vaccines, among other things. A few scientists have recognized both the importance of winning the trust of nonscientists and the challenges of doing so. Thus far, unfortunately, communication for the sake of trust is not a well-developed priority or skill set in the scientific community.

The challenge for scientists is all the greater because people tend to form trust and beliefs on the basis of the deep stories discussed in chapter 15. In general, people trust others when their own deep stories portray those others as trustworthy. People of common sense tend to be skeptical of claims about things that lie beyond their immediate, commonsense concerns. Though they can be persuaded to trust people and believe what they say, such persuasion is not easy, for any persuasive attempts will be interpreted through their deep stories. If the deep stories contain group and general biases, the challenge of winning trust will be all the harder. It will be possible only if intellectual and moral conversion takes place not only on the part of people of common sense but also on the part of the members of the scientific community. Intellectual and moral conversion of scientists is needed because, as a whole, they do not comprehend why people of common sense do not know or believe what science has revealed. Scientists do not even understand that what they say they "know" themselves is really a symbiotic fusion of the small amount of knowledge immanently generated by themselves individually within the vast amount of knowledge they have believed to be credible. Hence, the importance of trust in their work eludes their modes of communications, even on so urgent a matter. Real advance beyond these impasses of mutual mistrust between people of common sense and scientists therefore requires conversion on both sides. In particular, it requires "a few people who *seem reliable*" to both people of common sense and scientists alike, as Weart puts it. This crucial group of people are needed as go-betweens for the two camps; they can promote the kinds of genuine dialogue that can lead to such conversions.

One very remarkable person fostering such dialogue is Katharine Hayhoe. Hayhoe began her studies and early research in astrophysics and subsequently moved into the fields of atmospheric and climate science. She is currently the Chief Scientist for the Nature Conservancy and teaches at Texas Tech University. She lives in Lubbock, Texas, where her husband is pastor of an evangelical Christian church. She has been extremely active in speaking and writing to communicate her understanding and value of climate science to people like her neighbors and husband's parishioners, many of whom share the deep story examined by Hochschild.

Hayhoe does not begin her conversations with friends or audiences by listing the scientific facts regarding climate change. Her approach is summed up in a chapter title of one of her books: "Why Facts Matter—and Why They Are Not Enough." As she writes, "Bombarding someone with more data, facts, and science only engages their defenses, pushes them into self-justification, and leaves us more divided than when we began."[20] Instead she begins with things that unite her and those with whom she speaks. She asks them questions such as "Where do you live?

What or whom do you love? What activities do you enjoy doing? What do you do for work? Do you come from any particular culture, place, or faith tradition? And perhaps most importantly, what are you passionate about?"[21] She shares with them her own answers to such questions. Even though in "the U.S., white evangelicals are less worried about climate change than any other group,"[22] as an evangelical Christian herself, Hayhoe is able to establish relationships and build trust with people on those grounds. And only then does she explain how climate change threatens people and things precious to them. Her approach has been enormously successful.

If people of common sense are to engage in serious ethical reflection about the facts that science has revealed, they have to *believe* what scientists say that they know about matters of fact. In addition, serious ethical reflection on scientifically reported matters of fact will need to reflect upon the value claims by others as well. It is for this reason that Lonergan's method of functional specialties is so important in climate change and environmental ethics.

Chapter 18

Laudato si' and Integral Ecology

Introduction: Policy, Planning, and Implementation

All of the preceding chapters of this book have concentrated on what Lonergan called the functional specialties of History, Dialectic, and Foundations. Most of the substance of the book has relied on studies by historians of the environmental and climate sciences and the human responses, ethical and otherwise, to those developments. The functional specialties have been invoked at relevant places where their use illuminates and clarifies fundamental issues not explicitly addressed by the historians.

This chapter briefly takes up things that would fall under what Lonergan called the functional specialties of Policy, Planning, and Implementation.[1] In Lonergan's view of eight integrated functional specialties, these last three have important interconnections with the other five methodical specialties. That is to say, Policy, Planning, and Implementation depend upon what people take to be their foundations, and how they use those foundations to appropriate the ideas and actions of the past for the sake of the future. The arc of the functional specialties as an integral whole reveals how science and scholarship provide resources for ethical actions moving into the future.

This is not at all to say that people have not engaged in good policy formulation, planning, and implementation independently of Lonergan's method. Of course these activities have gone on for centuries. People have explicitly or tacitly adopted foundations and relied upon their fundamental assumptions and commitments to select what experiences were worth their attention and which questions about courses of action were worth pursuing. These fundamental commitments guided how they pursued innovative insights into possible courses of action and made judgments about which options were both realistic and worthwhile. Lonergan's method is intended to make these processes more self-reflective and thereby more deeply committed to the best and most objective values and actions. It is intended to alert policymakers, planners, and implementers to the long-term consequences that can be obscured by individual, group, and general biases as well as by unexamined philosophical counterpositions. It is intended to draw their attention to overlooked resources from the past that could make important differences in the future course of history.

In connection with the functional specialties of Policy, Planning, and Implementation, this chapter and the next will focus on two fairly recent developments of great importance in the history of climate change ethics: Pope Francis' encyclical *Laudato si'* and the United Nations climate change initiatives that began with the 1992 conference in Rio de Janeiro and reached an important culmination in its Sustainable Development Goals (UNSDGs).

The histories of these developments present difficulties for the approach in this book. First, because both of these developments are so recent, scholarly historical studies of the kind used in earlier chapters are not yet complete. Second, the actual historical processes set in motion by these developments are still ongoing. Third, their histories intertwine and influence each other in very important ways. Nevertheless, the history of *Laudato si'* and the UNSDGs is the continuation of the history surveyed in the preceding chapters. *Laudato si'* and the UNSDGs rely upon more than a century of the history of environmental and climate science. They draw upon the history of ethical thinking that reached a critical point in the uneven and incomplete transition from utilitarianism to holism. They grew out of the historical conflicts among those who affirm and those who deny the value of the environment as such, and conflicts among those who affirm and those who deny the reality of the human causes of climate change as documented by scientists. *Laudato si'* and the UNSDGs also incorporate several additional pathways, whose scholarly histories are still being written. These chapters, therefore, are more provisional than the previous chapters for these reasons. Nevertheless, both *Laudato si'* and the UNSDGs can both illustrate what Lonergan meant by Foundations, Policy, Planning, and Implementation, and how they can benefit from being viewed within in the larger heuristic framework of the eight functional specialties. This chapter is devoted to *Laudato si'* while discussion of the UNSDGs will be taken up in the next chapter.

Pope Francis and *Laudato si'*

For most of the world, Pope Francis' papal encyclical *Laudato si'* (2015) came as an enormous surprise. In style and content, it was dramatically different from almost every other major official document of the Roman Catholic Church. In style it differed from almost all prior papal encyclicals, which are ordinarily addressed to members of the Roman Catholic Church to exhort or clarify topics that the current pope finds of great importance to the church's members. By way of contrast, *Laudato si'* is explicitly and deliberately addressed to: "every person living on this planet." Rather than issuing instructions, Francis writes, "I would like to enter into dialogue with all people about our common home" (§§3, 14). Indeed the penultimate chapter is entirely given over to calls and guidelines for several new kinds of dialogue about how to meet the challenges posed by degradations of the environment and climate change. These dialogues, it is proposed, can lead to effective cooperative actions on many different levels (dialogue between science and religion, dialogues about economic and political matters at the international, national, local levels) all undertaken with transparency for the sake of protecting and promoting human dignity (§§163–201).

The encyclical is also different in content because of its comprehensive attention to the natural environment, its intrinsic goodness, the growing scientific knowledge of its complexities, the perils it faces, and the impacts of environmental degradation and climate change upon human beings, especially upon those most vulnerable: "The deterioration of the environment and of society affects the most vulnerable people on the planet" (§48). While some earlier papal encyclicals did mention the environment as an important component of Catholic social ethics (§§4–6), none has gone into such detail as does *Laudato si'*. This point is made emphatically by Seán McDonagh:

> Pope Francis is talking about the intrinsic rights of nature; that would've been unbelievable even 10 years ago—that the

> natural world has intrinsic value, not just instrumental value for human use. The World Council of Churches was way ahead of us in the 1980s, and we didn't fully cooperate with them. We were asked, actually, in 1990, to co-sponsor the World Convocation on Justice, Peace and the Integrity of Creation in Seoul, and the Vatican refused to do so. What Pope Francis has come up with, therefore, is quite revolutionary doctrine.[2]

In the previous chapters we have traced the dialectical history of the emerging awareness that there is need for a holistic approach to environmental and climate change ethics. *Laudato si'* is a very significant and influential contribution toward articulating this kind of holistic approach. The centerpiece, central chapter, and organizing principle of *Laudato si'* is what Pope Francis calls "integral ecology." It is a vision of "the universe as a whole" in which "human life is grounded in three fundamental and closely intertwined relationships: with God, with our neighbor and with the earth itself. According to the Bible, these three vital relationships have been broken, both outwardly and within us" (§66).

The encyclical's use of the word *integral* reflects its deliberate commitment to a holistic foundation. Its notion of an integral ecology is intended to provide a "framework"—a heuristic in Lonergan's terms—to guide collaborations among scientists and nonscientists as they seek "a better understanding of how different creatures relate to one another in making up the larger units which today we term 'ecosystems'" (§§132, 321, 140). The encyclical also emphasizes that the "ecology of daily life" (§147) with its economic, social, and cultural ecosystems are inescapably integrated and interconnected with environmental ecosystems within the whole of the integral ecology (§§138–55). The nonhuman world is not to be regarded as a mere "object to be used and abused without scruple," or which humans can treat with a "tyrannical anthropocentrism unconcerned for other creatures" (§§215, 68). Neither however are human beings alien invaders or parasites upon natural ecosystems, whose every action is nothing but an unjustifiable corruption of an otherwise pristine and separate realm. Just as it is natural for nonhuman living organisms to use and feed upon other living things in the integral ecology, it is also not unnatural for human beings to do likewise. That said, human interactions with and uses of the natural environment are compatible with it only if they are guided by "social love," which "is the key to authentic development: 'In order to make society more human, more worthy of the human person, love in social life—political, economic and cultural—must be given renewed value, becoming the constant and highest norm for all activity.' In this framework, along with the importance of little everyday gestures, social love moves us to devise larger strategies to halt environmental degradation and to encourage a 'culture of care' which permeates all of society" (§231). As Lonergan would put it, so long as human interactions with and uses of the other dimensions of the integral ecology are attentive, intelligent, reasonable, responsible, and loving, as long as they follow the norms of ethical intentionality, then they are legitimate continuations of the intelligible integral ecology of the emerging good.

Among other things, this integral framework requires a way of thinking about economic growth that is different from the "technocratic paradigm [that] tends to dominate economic and political life" (§109). There is urgent need for a "broader vision of reality," a "humanism capable of bringing together the different fields of knowledge, including economics, in the service of a more integral and integrating vision. Today, the analysis of environmental problems cannot be separated from the analysis of human, family, work-related and urban contexts" (§141).

For this reason the encyclical repeatedly speaks of sustainable and integral *development* as an essential dimension to integral ecology (§§13, 18, 50, 158–59, 167, 193). "Development" here means the transformation of social, economic, cultural, and interpersonal patterns of human living, undertaken in ways that intelligently build upon but do not destroy the physical, chemical, and biological ecosystems that make those developments possible in the first place. The encyclical claims that God created "a world in need of development" (§80)—but this means *integral* development rather than "anthropocentrically" distorted pathways of development (§69). The encyclical argues that protection of the environment is in fact "an integral part of the development process and cannot be considered in isolation from it" (§141).

Human and nonhuman ecosystems therefore are all integrated into "nature" rightly understood as a comprehending, integral whole (§139). The task of improving our understanding of the proper relationships of the various components of this whole requires the talents and cooperation of scientists and nonscientists alike, and the vision of an integral ecology offers a framework for achieving that cooperation.

The notion of an integral ecology functions similarly to what Lonergan calls an integral heuristic structure[3] in at least three ways. First, integral ecology is "heuristic" insofar as it guides research, dialogue, and action plans without predetermining the exact outcomes of thought and action. Second, it is "integral" insofar as it spotlights the complex systems of relationships among the many different natural and human constituents that make up the whole of the world (§23, 138). In particular, Pope Francis especially emphasizes that human beings and their social arrangements are integral components of the whole rather than alien to it. Third, like Lonergan, the encyclical envisions a whole that is dynamic and developmental, not finished and static.

The vision of *Laudato si'* is like that of Lonergan in yet another way. In Lonergan's account, scientific and ethical heuristic structures begin by defining problems and then devising processes that will lead to solutions.[4] *Laudato si'* follows a similar procedure. Its opening chapter begins with a "fresh analysis" of the current state of the environment, asking, "What Is Happening to Our Common Home?" (§17). It takes as a premise that the whole of nature and all its human and nonhuman interrelationships have intrinsic goodness, and that the intertwined human and natural ecosystems, and climate itself, together constitute "a common good." This common good, however, is under threat from pollution and climate change (§20–21), a throwaway culture (§22), consumerism (§34), depletion of water resources (§§27–31), the loss of biodiversity (§§32–42), excessive anthropocentrism (§§115–16), a practical relativism that reduces the value of other things and humans to their value only relative to oneself (§§121–22), a technocratic paradigm of unlimited growth (§§106–08), global inequalities (§§48–52), and the decline of human life and the breakdown of society (§43–47). This enumeration defines the complex set of problems that integral ecology is intended to address. The set of problems have resulted from the loss of a proper understanding of the interrelatedness of all human and natural ecosystems, and a proper valuation of that interrelatedness: "We cannot presume to heal our relationship with nature and the environment without healing all fundamental human relationships" (§119). Hence, the encyclical calls for dialogue and collaboration, and offers a heuristic vision of the goodness of the whole of integral ecology that can serve as a guide to the countless individual and institutional insights and actions that will be needed to solve the complex of problems set out in the first chapter of *Laudato si'*.

As leader of the Roman Catholic Church, Pope Francis also explains what the tradition of his particular religious body has to offer to the broader human dialogue about care for our common home. While endorsing many of the views of scientists and environmental activists, he says that his own tradition has something of importance to offer through its understanding of creation: "In the Judeo-Christian tradition, the word 'creation' has a broader meaning than 'nature,' for it has to do with God's loving plan in which every creature has its own value and significance" (§76, 84). Understood in this sense, creation means that human beings are enjoined to reverence and respect the value and significance of all creatures, and to make uses of natural entities always bearing their values in mind. On the other hand, creation also means that human beings are to be treated with their proper dignity (§81), something that has been repeatedly violated by actions whose impacts on the environment have had the worst consequences for the most vulnerable of human beings. It is the complex interrelationships between the Creator God, the natural ecosystems, and the human social and economic ecosystems that constitute the fullness of the integral ecology. As Pope Francis puts it, if the relationship to one of these three constituents is out of kilter, this will damage the relationships among the other components as well.

This book has endeavored to show that the dialectic of human thought and action reveals both the need for and the sporadic though imperfect emergence of a holistic foundation for environmental ethics. Clearly Pope Francis' integral ecology offers an important formulation of such a holistic approach.

Ecological Conversion

Pope Francis contends that both knowledge and commitment to achieving integral ecology requires an "ecological conversion." He quotes with approval the statement of the Australian Catholic bishops: "To achieve such reconciliation, we must examine our lives and acknowledge the ways in which we have harmed God's creation through our actions and our failure to act. We need to experience a conversion, or change of heart" (§218).

In an important essay, Neil Ormerod and Cristina Vanin have argued that Pope Francis' notion of ecological conversion is of such importance that it deserves "a more explanatory and analytic account of its full implications." They express concern that its "lack of precision" could "become a cipher into which various contents can be filled, contents which will not adequately allow humanity to avoid 'the catastrophe towards which it [is] moving.' "[5] In order to provide the desired explanatory clarifications, they draw upon Lonergan's account of three conversions—moral, intellectual, and religious—along with a fourth conversion ("psychic") proposed by Robert Doran. Their analysis is important for elucidating the notion not only of ecological conversion in *Laudato si'*, but also the conversion to holism that has marked the history of environmental ethics in general. Citing Lonergan, they describe conversion in general as "a radical shift of one's fundamental orientation, one's horizon."[6]

Prior to the present chapter, this book has drawn primarily upon Lonergan's notions of intellectual and moral conversion to address conflicts that have arisen in the history of environmental and climate change ethics and science. Lonergan also wrote at length about the third form of conversion—religious conversion—and religious conversion is also central to what Pope Francis means by "ecological conversion."

The phrase "religious conversion" by itself elicits a great many associations and reactions, many quite negative. Those negative reactions are quite understandable in light of the numerous and disgraceful abuses of religious ideas put to legitimating injustices toward humans as well as domination and exploitation of nature. In his famous essay "The Historical Roots of Our Ecological Crisis," Lynn White Jr. wrote, "By destroying pagan animism, Christianity made it possible to exploit nature in a mood of indifference to the feelings of natural objects."[7] While such historical realities cannot be denied, what Lonergan really meant by religious conversion is diametrically opposed to such distortions. Lonergan did not mean conversion to one or another identifiable religious organization. He did not mean adherence to some set of professed religious doctrines. While authentic religious conversion might eventually lead people to affirm doctrines or join religious communities, these are not what he meant by religious conversion itself.

Religious conversion is the decision to live in fidelity to unconditional love. It is a decision responding to the question "Will I live out the gift of . . . love, or will I hold back, turn away, withdraw?"[8] According to Lonergan, unconditional love dwells within the consciousness of every human being. Just as the unrestricted desire to know and to value permeates every human consciousness,[9] so too does unconditional love. Just as the unrestricted desire to know and value structures human thought and ethical deliberation by giving rise to questions for insights, judgments of fact and value, and decisions, so too the consciousness of unrestricted being in love "is a conscious dynamic state of love, joy, peace, that manifests itself in acts of kindness, goodness, fidelity, gentleness, and self-control."[10] Just as the unrestricted desire to know and value can escape notice unless practices of self-appropriation and discernment are developed, the same holds true for consciousness of unrestricted being in love.

According to Lonergan, unrestricted being in love is intrinsically related to the unrestricted desire to know and to value. "Just as unrestricted questioning is our capacity for self-transcendence, so being in love in an unrestricted fashion is the proper fulfilment of that capacity."[11] This means that intellectual and moral conversion are also intrinsically related to religious conversion. A commitment to unrestricted loving "places all other values in the light and the shadow of transcendent value. In the shadow, for transcendent value is supreme and incomparable. In the light, for transcendent value [of unrestricted loving] links itself to all other values to transform, magnify, glorify them." It is a love in which "the whole universe" along with "the human good becomes absorbed in the all-encompassing good."[12] Ormerod and Vanin go on to elaborate Lonergan's account of the wholeness of the good in terms of his "scale of values" pertaining to the human good[13] and its relationships to these three conversions.

Since unconditional love dwells within every human consciousness, religious conversion is a conversion to one's own most self, just as much as are intellectual and moral conversions. In addition, awareness of the unconditional love in one's consciousness, and the response of religious conversion both take time to mature, as do intellectual and moral conversion. One could say that intellectual and moral conversions remain incomplete until they are joined by religious conversion. The converse is also true. Living out the conversion to unrestricted being in love will remain incomplete until love opens up to all the questions about reality and goodness that are inspired by the unrestricted desire for knowledge and value. Again, openness to all questions (including scientific questions) about what is real and good extends also to questions about the ultimate reality of love as unrestricted. Conversely, radical openness to loving everything about everything cannot turn away from ever further questions about reality and goodness. The holism of the ecological conversions of Schweitzer, Muir, and Leopold incline in the direction of loving the entirety of the emerging universe and all life as beloved without limit.

Until this point in the present book this religious dimension of holism has not been brought into the discussion. But Pope Francis' encyclical raises the question of whether environmental ethics can be complete without considering the whole of emergent goodness as unconditionally beloved. Indeed, Ormerod and Vanin write, "Without religious conversion [to unrestricted loving] one cannot sustain a morally converted life."[14] Pope Francis does not call for conversion to Roman Catholicism. He calls for a loving embrace of the whole of natural and human goodness, and claims that without this, ecological conversion cannot be complete.

As the leader of the Roman Catholic Church, Pope Francis draws upon his religious tradition and identifies the ultimate of unconditional love as God and Creator. Over against all the misunderstandings, corruptions, and abuses of those terms, both Pope Francis and Lonergan find the ultimate foundational meaning from the consciousness of unconditional love: "An orientation to what is transcendent in lovableness . . . provides the primary and fundamental meaning of the name 'God.' "[15]

In §67 of *Laudato si'* Pope Francis seems to have White's essay very much in mind. He implicitly responds to White's criticism by explicitly rejecting any interpretation of Genesis 1:28 that would sanction "unbridled exploitation of nature by painting [humans] as domineering and destructive by nature." Yet Pope Francis also seems to pick up on the often overlooked concluding section to White's essay. There White argued that St. Francis of Assisi advanced a profound alternative orientation within Christianity. "Francis tried to depose man from his monarchy over creation and set up a democracy of all God's creatures."[16] Resonating with this theme, Pope Francis writes: "It is significant that the harmony which Saint Francis of Assisi experienced with all creatures was seen as a healing of that rupture" (§66). Jorge Bergoglio deliberately chose "Francis" as his papal name for exactly this reason. He does not expect all readers of the encyclical to accept the words adopted by the Catholic and other Christian traditions. Rather, he invites readers to view all natural and human reality and goodness as unconditionally beloved, calling forth loving responses.[17]

Laudato si' and Lonergan's Functional Specialties

Laudato si' exemplifies the basic thrust of Lonergan's last four functional specialties of Foundations, Policy, Planning, and Implementation,[18] although this might not seem obvious at first.

(1) The foundation for Pope Francis' *Laudato si'* is the notion of creation. For Catholics along with many other monotheists, creation is not an event that occurs in time. It is a relationship of loving dependence for the very existence and value of every finite thing upon an infinitely wise, capable, benevolent, and loving transcendent divinity. This is what Pope Francis says his religious tradition has to share with the rest of the world. The functional specialty of Foundations operates implicitly when he articulates his theology of creation in the language of integral ecology. This is a vision of the universe as a whole of three fundamental and closely intertwined relationships or interdependence among God, human beings, and the earth itself. Because creation is incomplete, these three relationships are dynamic and developing. They therefore require ongoing learning both in science and in human organization.

Pope Francis did not employ Lonergan's heuristic structures of emergent probability and the emerging good in articulating his foundation. Nevertheless, there are commonalities—the emphasis on the intrinsic goodness of the whole of creation, its complex dynamisms and incompleteness calling for further sustainable development, the explicit endorsement and call for the use of sciences in charting the path forward, among other things.

Pope Francis also brings to the fore the importance of explicit reference to the divine, transcendent creator. This is a most important dimension in Lonergan's work that has not been taken up in this book so far, since this has been primarily a philosophical effort. But as Lonergan puts it, if we raise the question about questioning—if we raise the question about why our questions for understanding, truth, and goodness actually do have answers—we are raising the question about God.[19] And since it is intelligent, reasonable, and ethical to raise such questions, they are natural continuations of the trajectory of the self-correcting processes of scientific and ethical questioning and answering.

Lonergan offers intricate answers to the question about why questions for insights have answers, concluding in his argument that reality is completely intelligible.[20] If this is the case, then there can be no "mere contingence,"[21] no merely brute facts in the universe. Thus Lonergan's account of emergent probability and the emerging good is incomplete unless one also faces these questions about the contingency of the natural and human universe. Lonergan argues that the contingency of the emerging universe and the contingency of its goodness point to an intelligent, benevolent, loving unrestricted intelligibility that people call by many names, but that Lonergan and Pope Francis call "God." For both Pope

Francis and Lonergan the ultimate foundation for ecological ethics is God. Both, however, are deeply committed to addressing ethical crises for all of humankind. Integral ecology and the emerging good provide foundations for dialogue across all religious and nonreligious peoples. They call for articulating religious foundations in terms that do not rely solely upon their specific religious doctrines.

(2) *Laudato si'* also articulates a number of policies in light of its foundation of integral ecology and on the basis of what both natural and human sciences (such as economics of poverty and development) have learned. Examples of these policies are: that creation as a whole and every creature should be treated as intrinsically valuable, especially that all actions should be taken to respect and promote human dignity; that special efforts should be taken to redress the disproportionate impacts of environmental and climate change damage upon the most vulnerable—the poor, women, and those otherwise marginalized; in particular, that "fossil fuels . . . need to be progressively replaced without delay" (§165); that countries should not "place their national interests above the global common good" (§169); and that new educational initiatives be undertaken "aimed at creating an 'ecological citizenship'" (§211).

(3) Perhaps the most sweeping of the policies articulated in *Laudato si'* is its call for dialogue at many different levels to accomplish planning and implementation in light of its foundation of integral ecology and its other policy guidelines. The encyclical is emphatic: "There are no uniform recipes, because each country or region has its own problems and limitations" (§180). Therefore, out of respect for the Catholic Social Teaching principle of subsidiary, the encyclical calls for the acceptance of plans made by experts and stakeholders at every level: global, regional, national, and local. It does provide pointers to some specific areas that collaborations on planning and implementation should address. It also cites approvingly, for example, the "action plan and a convention on biodiversity, and stated principles regarding forests," that were drawn up at the 1992 Rio Summit on Climate Change, though it also laments that these "accords have been poorly implemented" (§167). The encyclical further endorses other plans developed by other UN conferences, such as strategies for carbon credit exchanges and governance of the oceans. These are plans endorsed, but *not* devised by, Pope Francis or *Laudato si'*. They are endorsed because they are the plans and implementations that have originated from collaborations by many people and their prac-

tical, concrete creative insights and commitments to the wholeness of integral ecology.

The Path to *Laudato si'* and Dialectic

Controversies have arisen about some of the most important terms in the encyclical—especially *integral ecology* and *ecological conversion*. What do each of these terms mean? Among other things, ecology is the name of a field of science, like geology or biology. Is Pope Francis calling for a new science, and if so, what it its distinct field and methodology? There have also been calls for clarification of "ecological conversion." What kind of conversion is it? Is it a conversion to a group, like a religious or political conversion? Or is it a conversion of heart and mind that is not explicitly tied to an organization, such as Lonergan's intellectual, moral, and religious conversions? If the latter, is it completely different from or somehow related to Lonergan's conversions? How exactly have the backgrounds of Pope Francis and the other writers of *Laudato si'* influenced their understandings of integral ecology and ecological conversion? How have these influences affected the people and organizations who now attempt to implement the vision communicated by this encyclical?

It is not possible in this book to survey all of the existing historical studies of the influences on Pope Francis' *Laudato si'*. I will therefore limit myself to a study by Gerard Whelan. He analyzes the development of Pope Francis' theology, situating it in the broader movements of Roman Catholic ecclesiastical and theological history, as well as Argentinian and Latin American politics in the twentieth and early twenty-first centuries. The influence of Rafael Tello's thought was felt throughout Latin America in the latter half of the twentieth century, as it struggled through pervasive political and economic injustices. Whelan emphasizes how Tello's thought gave prominence to a "theology of the people."[22] By this is meant an approach to theology that carefully discerns how ordinary people are being led by their faith to respond to their concrete circumstances. Regarding Tello's influence on Pope Francis, Whelan writes: "Pope Francis stresses that all the baptized are the agents of evangelization. He suggests that notions of infallibility in matters of faith in Catholic theology should first and foremost be attributed to the whole people of God. Speaking of this infallibility, he describes it as related to the 'instinct of faith'—*sensus fidei*—which helps them to discern what is truly of God."[23] This theology of the people is prominent throughout *Laudato si'*, especially in its call for dialogue and collaboration with all people, whether Catholic or not, whether scientists or not.

Whelan explains that Tello saw the need for a more inductive Aristotelian approach over a "Platonic approach of applying universal systems of concepts to specific situations. Adopting a Thomist position, [Tello] insisted that the object of Christian care is 'not an abstract person, considered as some idea or conception,' but rather, 'the entire human person, in all his or her dimensions.'"[24] An inductive approach goes beyond simplified abstractions and is needed to apprehend the complexities of human beings in their concrete social settings. "Bergoglio's teachers explained that such a pastoral attitude implied adopting an inductive and historically conscious approach to theology, one that contrasted with the primarily deductive approach of the neo-Scholastic theology."[25]

Whelan characterizes Pope Francis' inductive theological approach with the popular phrase "See-Judge-Act." Francis picked up this notion from his teachers in the 1960s.[26] "See-Judge-Act" is a highly simplified way of speaking about structures of cognitional and ethical intentionality that Lonergan articulated in greater detail. This simplification certainly has its advantages in communication, but it also runs the risk of distortions. It omits the indispensable role of the self-correcting cycle of questions and insights that take place between seeing and judging, if judging is to be accurate and authentic. It also overlooks the essential roles played by further questions, refinement of feelings for values, and the nuancing of judgments that stand between seeing, and then judging and acting ethically. The details supplied by Lonergan are what make the difference between attentive, intelligent, reasonable, and responsible thinking and acting, on the one hand, versus what can be oversimplifications of the See-Judge-Act slogan, on the other. While *Laudato si'* is implicitly aware of this difference, the lack of explicit attention can at times obscure oversights and questions that are overlooked.

In addition to Tello, Romano Guardini also played a major role in Pope Francis' composition of *Laudato si'*. Years before his elevation to the papacy, Jorge Bergoglio went to the Jesuit Graduate School of Philosophy and Theology near Frankfurt, Sankt Georgen, Germany in 1986. There he intended to write a dissertation on the thought of the German theologian Guardini, although he had to abandon his studies after just a few months. Guardini is quoted explicitly in the body of *Laudato si'* (e.g., §203). Guardini's book *The End of the Modern World* is also referenced frequently in the encyclical's footnotes. Guardini saw "action" informed by theonomy as the way to overcome the philosophical and actual opposition between matter and spirit.

Whelan told me in a personal conversation that he suspected that Guardini's work may have introduced something that Lonergan would call a "counterposition" into Pope Francis' thought; this suspicion would have to be explored more fully using Lonergan's method. Whelan's book provides an extremely important historical and theological contextualization for approaching this question. However, scholars using Lonergan's method of functional specialties would have to rely on the future work of historians to trace in detail this development, and then to apply the method of Dialectic to clarify where that development is moving toward authentic wholeness of vision and where elements might have caused or are causing blockages, even in so capacious a vision as that presented in *Laudato si'*.

Chapter 19

The United Nations Strategic Goals for Sustainable Development

Pope Francis deliberately crafted and promulgated *Laudato si'* in hopes of influencing the upcoming United Nations climate summit (Conference of Parties 21 or COP21) in Paris in November 2015. Many have credited *Laudato si'* as indispensable for the widespread international adoption of the 2015 Paris Agreement. Nathaniel Rich said that "the most eloquent attempt to articulate a moral vision has come from Pope Francis,"[1] and it has been written that "Francis's pressure on COP21 was key to producing the Paris Agreement." Economist Jeffrey Sachs called *Laudato si'* "magnificent and breathtaking," especially in its proclamation that wealthy countries should pay their "ecological debt" to poor countries "by significantly limiting their consumption of nonrenewable energy and by assisting poorer countries to support policies and programs of sustainable development."[2]

The United Nations has played the leading role in advancing climate change ethics. Their task has been enormous—coordinating and reconciling over 190 nations to move toward addressing the ethical challenges posed by climate change science. The UN process reaches back to its Conference on the Human Environment in 1972, which established the United Nations Environmental Program (UNEP). The UNEP later collaborated with the World Meteorological Organization (WMO) to form the IPCC in 1988. The process moved through annual Climate Summits (COPs) beginning with the Rio Conference in 1992 and with major achievements in Kyoto (1997), Paris (2015), and Glasgow (2021).

The accords agreed upon at these COPs by over 190 nations have been major steps toward realizing environmental and climate change ethical values.

Activists and protest demonstrators have become increasingly impatient with this process because the emission of GHGs into the atmosphere has not only continued but accelerated over the twenty-seven years since the Kyoto agreements. The intransigence, denial, and self-centeredness that have contributed to these continuing increases are indeed reprehensible and climate science denial has played a major role in slowing responses. Nevertheless, it must also be borne in mind that meeting the massive ethical challenges posed by climate change science can be compared with trying to turn around the *Titanic* upon first sight of the iceberg. The *Titanic* of climate change began with the onset of the Industrial Revolution around 1750. Climate change science only began around 1880 and a scientific consensus about global warming and the human causation of it began to form only around 1977. Even by the late 1980s there were significant hesitations by numerous scientists about the degree of probability of scientific predictions. In the meantime, countless human habits along with economic and political institutional patterns have been building for more than two hundred years. It is not easy to turn around two hundred years of global historical momentum. It is not like throwing a single circuit breaker. This is not to excuse the ethical failures, but it is to point out that ethics has to respond to real situations and to come up with insights into the intelligent steps needed. Climate change ethics cannot ignore the numerous and complex challenges involved in turning around the hearts and minds of the world's population and coming up with insights into millions of concrete solutions.

The UN process has been more massive and complex than anything discussed previously in this book. It has been gigantic because over 190 nations and thousands of NGOs have been involved in the process. Over the course of the process, ever more topics have been incorporated, especially as the intricate connections between global economic inequality and GHG reductions have come to be understood. Agreements have been reached over ever-increasing numbers of issues and about ever more numerous concrete steps to be taken. Various organizations have been formed to meet technological, economic, and political challenges in implementation. Countless insights and initiatives have flowed from these organizations.

For many this has been too slow, and some of the wealthiest nations have been justifiably criticized as doing far too little. While this is all quite true, exclusive focus on what has not been done has obscured how much has been done by tens of thousands of unsung people who have contributed to progress so far.

Because of the magnitude and complexity of the UN process, because of the numerous conflicts that have permeated it, and because it is still ongoing, no comprehensive history of the process has been completed so far. There are numerous partial historical studies, however, including the studies referenced in chapter 15 about the US' taciturn involvement with the Rio and Paris COPs. This book cannot do justice to all of these partial studies; such a task would require a second book comparable to the present one, in order to chart the history of the UN process since 1992. That said, this book would be woefully incomplete without some commentary on the deeply important role of the United Nations process in environmental and climate change ethics. For that I will be relying chiefly on the work of Jeffrey Sachs, especially his *Age of Sustainable Development.*[3]

Jeffrey Sachs on Sustainable Development

A holistic ethical vision comparable to that of *Laudato si'* (and the heuristic of the emerging good) was presented in the United Nations' adoption of the Sustainable Development Goals (UNSDGs) at the Rio+20 summit in June 2012 and the Paris Agreement of 2015. The holistic vision of the UNSDGs is formulated in four layers: (1) At the apex, there are four interrelated dynamic systems that must be taken into account in any ethical response. (2) The analysis of these interrelated systems led to formulation of the seventeen Sustainable Development Goals themselves. (3) Under the UNSDGs there is a third level of 169 more detailed targets or implementation strategies to be adopted by local governments under the guidance of the two previous levels of the heuristic of the UNSDGs.[4] These targets are accompanied by 232 "indicators" to help measure progress in achieving the targets. (4) Finally, there are concrete actions to enact these targets, which are to be implemented at local and personal levels; here the need for appropriate systems of governance is especially important. These four levels of the UNSDG program correspond remarkably with Lonergan's functional specialties of Foundations, Policy, Planning, and Implementation.

(1) First, then, at the heart of the SDGs is the holistic framework consisting of "not just one but four complex [nonlinear] interacting systems" or "pillars" of sustainable development: the earth's environment, the global economy, social systems, and governance.[5] Sachs, who was instrumental in the formulation of the UNSDGs, put it as follows: "The fundamental question is how to take our knowledge of the interconnections of the economy, society, the environment, and governance and apply it to determine how to produce prosperous, inclusive, sustainable, and well-governed societies, that is, how do we achieve the SDGs?"[6] The following three features are essential to this interlocking network of dynamic systems.

(i) Each of the four systems are complex and "nonlinear"—they are what used to be called "chaotic" but now specialists prefer to call them complex, dynamic systems. As we saw in chapter 13, dynamic systems involve complex feedback mechanisms, are highly sensitive and even volatile to minute changes in conditions, and for these reasons precise outcomes are intrinsically unpredictable. Yet this does not preclude thinking in terms of probable outcomes over large enough populations and spans of time.

(ii) Because the dynamic systems are interdependent, solutions to any one problem area—such as mitigation of disastrous climate changes—inescapably require significant corresponding modifications in the other three dimensions. Again, any proposals for modifying the global economy in order to mitigate climate change cannot ignore the impacts and interdependencies on dynamic systems of social interaction and governance.

(iii) Because of the dynamic reality of all four systems, the goals they imply must be *developmental*, not static, unchanging universal ethical imperatives.

These three features mean that it is impossible to merely *preserve* the environment and climate just as it is without modifying it. The Paris Agreement signatories affirmed that natural ecosystems and climate systems are themselves already dynamic, self-modifying, and nonlinear. Hence there is need for an ethics of dynamic development. This does not mean that the UN endorses indiscriminate development of just any kind. The SDGs certainly do not endorse the prevailing paradigms of economic development for the sake of maximizing financial profits nor just any kind of technological innovation. Sachs explains in detail the insufficiency of growth in Gross Domestic Product (GDP) as the proper measure of social well-being. Growth in GDP must be supplemented by additional indices such as Purchasing Power Parity (PPP, i.e., a measure of wealth distribution) and the Human Development Index (HDI), "life satisfaction," social capital, trust, generosity, and availability of physical and mental health services.[7]

Sustainable development means discerning how the natural and human environment can develop in ways that are not mutually destructive. This does not mean refraining from every kind of human change; it means undertaking ethical changes that will sustain conditions for conjoined human and nonhuman development that is intelligible.

Hence, the UNSGD emphasis has been on discerning what conditions make sustainable development possible. Rachel Carson has observed: "It took hundreds of millions of years to produce the life that now inhabits the earth—eons of time in which that developing and evolving and diversifying life reached a state of adjustment and balance with its surroundings. . . . Given time—time not in years but in millennia—life adjusts and balance is reached. For time is the essential ingredient; but in the modern world there is no time. The rapidity of change and the speed with which new situations are created follow the impetuous and heedless pace of man rather than the deliberate pace of nature."[8]

In other words, dynamic evolution is the very essence of nature, but evolution has had a relatively gradual pace (the "mass extinctions" excepted). This means that developmental changes occur slowly enough for ecosystems to evolve and replace one another as the underlying conditions change. It takes both scientific analysis and heuristic guidance of ethical decision-making and conversions to achieve sustainable development.

Lonergan's foundational concepts from his heuristic of emergent probability and the emerging good offer important insights here. He defined authentic development as an *intelligible* "linked sequence" of dynamic, self-modifying systems, in which the operations of the previous system set the conditions for the emergence of the subsequent ones.[9] Lonergan's definition of development excludes sequences of actions that would destroy, rather than promote, the conditions for the emergence of new intelligible, sustainable, and more nuanced dynamic systems, whether natural, economic, social, or governmental.

(2) Responding to the four complex, dynamic, and interconnected systems required the articulation of heuristic goals for a dynamic ethics

of sustainable development. There are seventeen such goals, and perhaps surprisingly, the 2012 United Nations Conference on Sustainable Development (aka Rio+20) declared: "Eradicating poverty is the greatest global challenge facing the world today and an indispensable requirement for sustainable development." To explain poverty's prime position as a sustainable development goal, Sachs situates the seventeen SDGs in the context of the history of the waves of economic development dating to the Industrial Revolution (ca. 1750). "Until the Industrial Revolution, most of the world was poor and rural, and so the gaps of rich and poor countries were quite narrow, not like the huge gaps today."[10] Sachs also shows how the Industrial Revolution made possible dramatic improvements in agricultural output, health care, mortality rates, transportation, education, and standards of living. Since the Industrial Revolution the global population has increased by about seven billion people (from approximately 800 million in 1750 to 7.8 billion in 2020), but Sachs is most emphatic that the wealth generated by these centuries of economic development has not been distributed at all equitably. The same shift in technology and organization underlies both climate change and the current situation of global disparity between rich and poor nations.

The economic developments that began with the Industrial Revolution are needed to support just about 90 percent of today's global population. The simplistic answer to addressing global warming and climate change—reverting to the 1750 economic and technological state—would result in the loss of billions of human lives. The picture becomes even more complex as Sachs also presents the research showing that birth rates decrease with the increased education of women. These and other factors that Sachs analyzes imply that a number of dynamically interacting factors of both natural and human environments need to be addressed holistically. These interconnections were assembled into the seventeen UNSDGs as follow:

1. No Poverty
    - End poverty everywhere, in all its forms.
2. Zero Hunger
    - End hunger, achieve food security and improved nutrition, and promote sustainable agriculture.

3. Good Health and Well-Being

    - Ensure healthy lives and promote well-being for all at all ages.

4. Quality Education

    - Ensure inclusive and equitable quality education and promote learning opportunities for all.

5. Gender Equality

    - Achieve gender equality and empower all women and girls.

6. Clean Water and Sanitation

    - Ensure availability and sustainable management of water and sanitation for all.

7. Affordable and Clean Energy

    - Ensure access to affordable, reliable, sustainable, and modern energy for all.

8. Decent Work and Economic Growth

    - Promote sustained, inclusive, and sustainable economic growth, full and productive employment, and decent work for all.

9. Industry, Innovation, and Infrastructure

    - Build resilient infrastructure, promote inclusive and sustainable industrialization, and foster innovation.

10. Reduced Inequalities

    - Reduce inequality within and among countries.

11. Sustainable Cities and Communities

    - Make cities and human settlements inclusive, safe, resilient, and sustainable.

12. Responsible Consumption and Production
    - Ensure sustainable consumption and production patterns.
13. Climate Action
    - Take urgent action to combat climate change and its impacts.
14. Life below Water
    - Conserve and sustainably use the oceans, seas, and marine resources for sustainable development.
15. Life on Land
    - Protect, restore, and promote sustainable use of terrestrial ecosystems, sustainably manage forests, combat desertification, halt and reverse land degradation, and halt biodiversity loss.
16. Peace, Justice, and Strong Institutions
    - Promote peaceful and inclusive societies for sustainable development, provide access to justice for all, and build effective, accountable, and inclusive institutions at all levels.
17. Partnerships for the Goals
    - Strengthen the means of implementation and revitalize the global partnership for sustainable development.[11]

Taken together as a whole, these SDGs constitute a rather highly articulated heuristic structure of policies and ethical imperatives.

(3) Under the guidance of the previous two levels of the heuristic, the signatories to the 2015 Paris Agreement promised at their national and local levels to work toward meeting 169 targets using "indicators" to measure progress toward the targets. More recently, agreements at the Glasgow COP26 conference have included targets and indicators that articulate plans even more fully in order to achieve the seventeen SDGs.

In order to exemplify this heuristically structured process, I will focus in more detail upon the five targets defined by the UN under

SDG 7, "Ensure access to affordable, reliable, sustainable, and modern energy for all." They are:

- Universal access to modern energy (e.g., by 2030 ensure universal access to affordable, reliable, and modern energy services and clean fuels for cooking).
- Increase global percentage of renewable energy.
- Double the improvement in energy efficiency.
- Promote access, technology, and investments in clean energy.
- Expand and upgrade energy services for developing countries.[12]

These five targets are accompanied by six "indicators" in order to assess progress toward the targets. These include proportions of: a local population with access to electricity (and the stated objective of universal access by 2030, especially in underdeveloped economies); populations with access to clean fuels for cooking; renewable versus nonrenewable energy sources; and support by developed countries for increasing renewable energy in developing countries.

(4) It is beyond the scope of this book to engage all the reports of efforts to implement the UNSDGs and targets. These implementations are so recent that critical historical accounts cannot yet be written, but one ongoing source is the UN website for the SDGs, which is regularly updated to track the actions being taken and progress being made (or not) toward implementing the goals and targets. In 2021 the website reported that 5,475 actions and 3,086 events (and 1,303 publications) had been undertaken toward these ends. While reporting where progress has been made, these updates also increasingly focus on worldwide failures and the distances between agreed upon goals and targets and actual implementations. For example, the website reported that the COVID-19 pandemic led to the first rise in global extreme poverty in a generation.[13]

In place of the full historical and dialectical analysis that will eventually be needed, I again draw upon the work of Sachs, who offers some of the most important examples of how implementation has and can take place. Here I will focus upon his account of "differential diagnosis" as a method for implementing the UNSDGs.

Because of the complexities of the four interacting dynamic systems, Sachs emphasizes that one size cannot fit all: "There seems to be a misguided desire for overly simple explanations of complex economic dynamics." Ethical responses to climate change therefore have to involve what he calls "differential diagnosis" modeled on the "art of modern medicine [that] has advanced beyond pronouncing a single cause of a disease . . . or prescription."[14] As with all genuinely ethical deliberations and actions, Sachs echoes the claim of the ethics of discernment that the right courses of action begin by addressing the specific questions that arise out of concrete circumstances.

Sachs originally developed his method of differential diagnosis within the UN framework that preceded the SDGs, namely the Millennium Development Goals (MDGs), which were focused on the reduction of extreme poverty. Sachs argues that, like modern medical practitioners, modern economists should make specific differential diagnoses of poverty and other economic woes that are "accurate and effective for the conditions, history, geography, culture and economic structure of the country in question." Along these lines, Sachs compiled a diagnostic "checklist" of seven interconnected factors that prevent nations from emerging out of poverty. He applied the checklist to a variety of nations, showing how it can help to identify the problems specific to each nation, and the particular measures that can be adopted to overcome them.[15]

Although threats to planetary warming are global rather than local, climate changes themselves vary according to geographic terrains as well as to economic, social, and political circumstances. Hence, differential diagnosis of each of the dynamic systems at local levels is likewise necessary to bring about appropriate local changes to meet the holistic UNSDGs. Sachs goes on to explain how differential diagnosis has been applied in implementing projects to meet the UNSDGs, such as the Millennium Villages Project (MVP) and the Sustainable Development Solutions Network (SDSN).[16]

The SDSN coordinated cooperation among many different institutions as is needed to meet the ethical challenges of climate change—which, in turn, impact and are impacted by the three other complex dynamic systems of global economy, social inclusion, and authentic governance. Individual efforts can be effective only if conjoined so as to contribute in an intelligible fashion to these holistic goals. It is noteworthy, however, that there are no UN enforcement mechanisms to ensure that local plans are made or implemented responsibly, nor are there measures designed to punish bad actors. Instead, the annual COPs call upon the signatory nations each to present accounts of how well they have done in meeting their own planned targets. This is why a dynamic heuristic ethical vision, and the conversions it implies, are irreplaceable. Force alone cannot achieve the seventeen UNSDGs. In the end, the process cannot be realized unless those converted to wholeness cooperatively follow their shared call and reach even those resistant to that call.

The UNSDGs and the Functional Specialties

I have proposed that Lonergan's method of functional specialties supports and clarifies the UNSDGs, but how do they do so? Lonergan did not intend for his method of functional specialties to be either something completely unprecedented or something that would completely replace the policy formulation, planning, and implementation that people have been doing throughout human history. His intent, rather, was to make those activities more self-reflective and more methodical and therefore more effective in bringing about what is truly valuable and good. The method is meant to enhance those processes and gradually to reverse the oversights, biases, and counterpositions that disrupt them. The distinctiveness of Policies, Planning, and Implementation precisely as functional specialties has three features: (1) They are not actual policies, plans, or implementation but rather heuristics to guide those processes toward forming policies, plans, and implementations. (2) They are done self-reflectively with clear awareness of and commitment to the normativity of the structure of ethical intentionality (i.e., doing these operations on the basis of what Lonergan called "self-appropriation"). As such, they can make people more aware of the interferences being caused by biases and counterpositions. (3) Policies, Planning, and Implementation are explicitly carried out on the basis of findings from the prior five functional specialties, and especially History, Dialectic, and Foundations.

The connections between Lonergan's functional specialties of Foundations, Policy, Planning, and Implementation are perhaps most apparent in the United Nations' SDGs, although it is highly unlikely anyone working in this area had any awareness of Lonergan's thought.

(1) Lonergan's Foundations situates the four complex dynamic systems of the UNSDGs within the heuristic frameworks of emergent probability and the emerging good. Chapter 7 outlined the Foundational

heuristic structures of emergent probability and the emerging good that Lonergan developed to articulate the implications of intellectual and moral conversion. Both structures emphasize nonsystematic, dynamic emergence as the central feature of their Foundational claims. In parallel fashion, the UNSDGs rest on the foundation of the interconnections among the four complex dynamic systems of the earth's environment, the global economy, social inclusion, and governance. Lonergan situated these three human dynamics (economic, social, and political) as dimensions of the good of order in his structure of the human good. He argued that they emerge from the prior levels of the emerging good pertaining to physical, chemical, geological, meteorological, biological, and animal-psychological schemes of recurrence. While the UNSDGs provide greater specificity, Lonergan's heuristics provide a more comprehensive heuristic of the whole to be taken into account in Policy, Planning, and Implementation. At some point in the future, questions will have to be faced about how this larger integral framework might concretely enhance the UNSDGs themselves, as well as their implementations.

(2) The functional specialty of Policy is tasked with making value judgments and decisions about which values are to be affirmed and in what order of priority. In doing so, it relies upon the guidance of the Foundational categories, as well as the studies that have been done in the earlier functional specialties including History. The formulation of the seventeen UNSDGs themselves did exactly this, without explicit reference to the structures that were being invoked implicitly. The lengthy debates and inputs regarding ever further questions on the part of dedicated and authentic experts were essential to the formulation of these seventeen goals as policies.

(3) While the Foundations and Policies of the seventeen UNSDGs were adopted globally, Planning required the formation of localized efforts, primarily at national levels, although NGOs and some corporations have also joined in planning efforts to realize the goals. As Lonergan remarked, because real and ethical solutions require insights that occur to people "on the spot, their only satisfactory expression is their implementation, their only adequate correction is the emergence of further insights" to people familiar with local circumstances.[17] Local environmental, social, economic, and political schemes of recurrence vary so widely across the planet that it would have been both foolish and irresponsible for the UN to dictate some universal set of plans to

be followed by every nation. There was great ethical wisdom in leaving it to nationals to formulate their own specific goals and plans. There was also political prudence involved, since it is unlikely that any nations would have approved the UNSDGs without that latitude. Hence, the Rio Conference agreed to 169 targets under the various UNSDGs, but responsibility for the concrete planning needed to meet those targets was left to individual nations. While group biases no doubt influenced the commitment to delegate plans to national levels, even behind these biases stands the reluctance to have one's local understanding superseded by a remote central authority that could not possibly have the insights and valuations found in local populations.

(4) Since the concrete planning needed to realize the seventeen UNSDGs was ceded to nations, implementation fell to them as well. Wisely, however, Rio also committed the signatories to participation in the annual Conferences of the Parties (COPs). Each year the nations come together to report on their progress toward the goals and plans to which they have committed themselves. The body of signatories hold each other responsible. These COPs also become occasions when new information is exchanged (especially from the IPCC reports) and new technologies are explained that can be used to modify targets and make for new cooperative agreements and collaborative efforts. For example, the following are some of the decisions that came out of the 2021 COP26 in Glasgow:

- A commitment was made by all signatories to "phase down" unabated coal power and inefficient fossil fuel subsidies.[18]
- Thirty-four countries along with several banks and financial agencies pledged to terminate international funding for an unabated fossil fuel energy sector by the end of 2022.
- More than forty countries pledged to move away from coal.
- The US and the EU announced a global partnership to cut emissions of methane by 30 percent by 2030, with more than one hundred countries joining the agreement.
- Governments of twenty-four developed countries and a group of major auto manufacturers committed to working toward all new cars and vans being zero emissions globally by 2040.

- Forty-five countries pledged to provide more than $4 billion for transitioning to sustainable agriculture.
- More than one hundred world leaders promised to end and reverse deforestation by 2030.
- The pact pledged to significantly increase money to help poor countries cope with the effects of climate change and make the switch to clean energy.
- Financial organizations controlling $130 trillion agreed to back clean energy technologies and renewable energy, as well as direct finance away from fossil fuel–burning industries.
- However, the goal of $100 billion/year to aid developing countries, which was agreed to at the Paris COP21 in 2015, is still far from being realized.

These are just a few examples of the agreements reached at COP26.[19] These agreements will become the basis upon which the signatories will hold each other accountable for implementations at future COP summits. They will also be the bases for activists to continue to pressure local governments and businesses.

Conclusion

Lonergan did not invent something completely new in formulating his integrated method of eight functional specialties. Because they are based upon the structures of cognition and ethical intentionality that every human being uses repeatedly day in and day out, it is to be expected that the patterns he formalized in his work would be found in spontaneous efforts such as those embodied in *Laudato si'* and the UNSDG processes. Lonergan's method is meant to enhance those considerable and admirable efforts with the addition of attentive and careful self-appropriation of what goes on as people engage in policy formation, planning, and implementation. That explicit self-appropriation can make each person's work more effective and can accelerate genuine interdisciplinary collaboration. While originally developed to meet the needs of a modern theology, Lonergan recognized that his method of eight functional specialties could enhance the ethical authenticity of individual research and collaboration across disciplines such as climate change science, social sciences, and policy formation at governmental and corporate levels. This extension is possible because the method is grounded in self-appropriation of cognitional and ethical structures that already operate (imperfectly) in the thinking and acting of every human being. This chapter has offered concrete illustrations of how Lonergan's functional specialties of Foundations, Doctrines, Planning, and Implementation can be related to the real-world work of the United Nations as it formulated and has moved toward implementing the policies of its seventeen UNSDGs.

Conclusion

Environmental ethics for our planet earth must be holistic. Because earth is an exceptionally complex and dynamically evolving whole, environmental ethics must be correspondingly complex and dynamic. This means that environmental ethics has and will continue to come about through the coordinated actions of all people on the planet earth. This book has been dedicated to making a strong case for these points.

A dynamic and holistic ethics cannot consist in a single, simplistic imperative, nor even a small number of such imperatives. Ethics for the planet requires contributions from every human citizen of the planet. This means that each person's contributions will differ from those of others and yet must be coordinated with the actions of others in complex and evolving ways. This requires a framework within which each person can discern her or his contributions and comprehend in at least a general way how they connect with those of others. In this book I have drawn upon the work of Bernard Lonergan and my own extension of his work into an "ethics of discernment" and its accompanying heuristic of the emerging good to offer a framework that I believe goes a long way toward providing such guidance.

One of the greatest challenges for environmental ethics has been posed by the ongoing developments in environmental and climate change sciences. These developments have dramatically changed how human beings understand the earth. But the sciences alone do not reach ethical insights or value judgments or decisions about what actions ought to be taken in light of the knowledge presented by scientific research. This is why the first chapter took up the challenge set forth by David Hume regarding the proper relationships between knowledge of facts and knowledge of what ought to be done. While I agreed with Hume that statements of "ought" cannot be validly derived by deductive logic from statements of fact, I disagreed emphatically with the path he took from that observation. I argued there (and in my previous book) that neither scientific nor ethical reasoning can be reduced to formal logical reasoning. Rather, both scientific and ethical reasoning are matters of what Lonergan called the "self-correcting cycle" of experiences giving rise to questions, questions pursuing and arriving eventually at insights, insights stimulating search for additional experiences and further questions, resulting in insights that modify, correct, or even abandon earlier insights. The methods of the sciences, including environmental and climate change sciences, are specialized elaborations of this basic pattern of reasoning, as I have endeavored to show throughout the chapters of this book.

Ethical reasoning in general also exhibits this same self-correcting cycle and is likewise adapted and specialized according to the ethical matters under consideration. I also agreed with Hume that feelings enter into ethics, but I disagreed with how he construed their role. For Hume, feelings motivate actions, while reasoning cannot do so. Moral statements, according to Hume, are simply expressions of passions, and moral reasoning is no more than efforts to shape some passions (such as sympathy) in subordination to other passions (e.g., desire for profit). Moral reasoning for him, therefore, is no more than a "slave to passions."

I disagree with him that reasoning is the slave to feelings. Feelings give us an indispensable consciousness of values, but feelings alone do not automatically provide objectively correct evaluations of the situations out of which they arise, nor do they automatically produce actions. Feelings provide a provisional assessment of the values inherent in contemporary situations, but feelings are not infallible. Feelings also offer initial valuations of the possible courses of action suggested by insights. But when the self-correcting process follows its proper course by considering further questions and insights about those feelings and into alternative courses of actions, then feelings and reasoning together can arrive at objective ethical responses to what is known about facts, scientifically or otherwise.

Because along with Lonergan I have stressed the great importance of historical thinking to ethics, I have relied upon the scholarship of many historians of both science and ethical thought. While they knew nothing of each other's work, historian Spencer Weart provided a strong endorsement for Lonergan's emphasis when he wrote that only by "following how scientists in the past fought their way through the uncertainties of climate change, [can we] judge why they speak as they do today."[1] The same can be said of the importance of understanding the histories of environmental science and environmental ethics.

There are a great many other important historical studies of both areas that are not discussed explicitly in this book. That is because this book is not a historical study itself. It is, rather, an effort to add ethical thought to historical scholarship by employing the larger framework within which Lonergan situated the importance of historical scholarship—namely, his method that integrates historical scholarship with the contributions of seven additional "functional specialties."

In part 1, "Environmental Ethics," I synthesized the works of historians of environmental science and ethics, and showed that they revealed the need for a holistic approach—even a "conversion"—in environmental ethical thought and action. I went beyond those historical studies to show how attempts to provide that holistic approach were beset with a variety of difficulties and performative contradictions of the sort that Lonergan would characterize as "dialectical." I then explained how my holistic framework of the "emerging good" (inspired by Lonergan's work) can resolve those issues, and that this approach is grounded in a "self-appropriation" of the structure of human ethical intentionality. I also showed how adopting this framework has strong resonances with the holistic environmental "conversions" that were self-described by such luminaries as Albert Schweitzer, John Muir, and Aldo Leopold. The structure of ethical intentionality is not something Lonergan or I invented, but something to be discerned as already immanent and operating in the consciousness of all human beings. Hence, it is a framework for a holistic ethics available for people of all cultures as a basis for collaboration on environmental issues. It also provides a standard for criticizing the shortsighted and self-serving proposals of those who prefer to ignore the call of "further pertinent questions" that do arise in the self-correcting process of ethical reasoning about our environments.

In part 2, "The Ethics of Climate Change," I relied on additional scholarly historical studies that traced, first, the century-long and complex development of climate change science, and then the conflicting, dialectical reactions to that emerging body of scientific knowledge. Matters are made difficult by the fact that the physical processes of human-caused global warming and climate change themselves had been advancing for two centuries before a scientific consensus about these factual matters of fact began to mature. For two centuries many worldwide economic, social, and political practices, assumptions, and institutions had developed that are ill-suited to responding properly to the realities of climate change. Even more so than is the case with environmental ethics, climate change ethics calls for a significant shift or "conversion" in some of the most basic assumptions about climate and human practices that were constantly shaped and reinforced over the past two centuries. In part 2 I relied on historians to show how climate change ethics has reached a crisis of denial that is still impeding fully ethical responses to the consequences of ignoring climate change science. I argued that what Lonergan called a "moral conversion" to a holistic ethical framework is necessary. In addition, there is need for an "intellectual conversion" that better recognizes the kind of knowledge and certitude that has been achieved over the history of climate change science. Toward the end of part 2 I show how two achievements, *Laudato si'* and the UN Strategic Development Goals, have made significant contributions to meeting the challenges of environmental and climate change ethics—and that these achievements resonate with Lonergan's ethical method of functional specialties.

Environmental and climate change ethics is not merely a set of prescriptions that ought to be followed. It is a massive and complex set of human actions that are needed to respond collectively to our complex world and its challenges. The work has begun on the part of millions of people. Already millions have become "ecologically converted" as Pope Francis has called it. Already these many have come up with new insights and value commitments about how to heal and improve our wounded world. Yet many obstacles remain to be overcome. Those obstacles might be overcome temporarily by governmental power, and sometimes this will be a last resort. But governmental power will not go to the root of the problem. The problem ultimately requires conversion of human hearts and minds to an ethical process that takes seriously feelings for the entire range of values about things natural and human, and that respects the need to answer all questions—scientific, cultural, and ethical—about what is to be done.

Appendix

A Method for Environmental Ethics

1. From *Method in Theology* to Method in Ethics

In the last two chapters of *The Ethics of Discernment*[1] I argued that the proper method for ethics is the eight functional specialties that Lonergan worked out as the method for theology. I have structured *Toward Environmental Wholeness* along the lines of these eight functional specialties. I have deferred a detailed account of the method itself to this appendix.

In Lonergan's method of theology, there are four functional specialties for accurately understanding and evaluating the past: Research, Interpretation, History, and Dialectic. There are four more functional specialties that are designed to facilitate individual and group decisions for moving forward: Foundations, Doctrines, Systematics, and Communications. In the context of philosophical ethics rather than theology, the last three would be modified into Policy, Planning, and Implementation, following Lonergan's own suggestion.[2]

The eight functional specialties were developed on the basis of the key components in Lonergan's structures of cognition and ethical intentionality. Therefore, the next section briefly surveys his accounts of those structures before turning to an exposition of the functional specialties themselves.

2. Cognitional Structure, the Structure of Ethical Intentionality, and Self-Appropriation

This method of eight integrated functional specialties is a sophisticated refinement of what ordinary people do spontaneously and what trained scholars do in more disciplined ways. Everyone receives input about ethical matters from a variety of sources: family, neighbors, teachers, clergy, friends, news and social media, cultural legends and stories, and so on. Ordinary people spontaneously sort out from among this vast input what they will take to heart and what they will leave aside. The ways they make such determinations can be fairly consistent or rather haphazard. Scholars have developed methods to make for a more critical evaluation of ethical claims from the past and, to some extent, offer more critical advice about how to proceed toward the future.

It took Lonergan a long time to develop his more sophisticated refinement of receiving the past ("mediating phase") and addressing the future ("mediated phase").[3] The basis of his approach was what he called "self-appropriation": deliberate and reflective attentiveness, understanding, and explicit commitment to the basic structures that ordinary people use spontaneously and scholars have refined with varying degrees of self-awareness. In *Insight* and later in his seminal essay "Cognitional Structure," Lonergan set forth the basic discoveries from his own practices of self-appropriation. Most fundamental among these were his discovery that "human knowing is a dynamic structure [that] implies that human knowing is not some single operation or activity, but on the contrary, a whole whose parts are cognitional activities."[4] According to him, most people assume that seeing or touching or something similar is the "single operation or activity" that constitutes human knowing. This is most evident in empiricist or positivist epistemologies. Because they privilege seeing or touching as the beginning and end of all that deserves to be called "knowledge," they therefore hold that reality is no more than what is visible or tangible, that is, that reality is material. Everything else is a projection of the human mind onto the bare bones of material reality. Post-empiricist thinkers like Kant recognized the greater complexities in human thinking that go beyond sensation, but they did not conclude that these complexities lead to knowledge of reality. As Lonergan put it, idealism is "unable to break completely" with the notion of knowledge of reality that takes looking and touching as its paradigm.[5]

By way of contrast, Lonergan's practice of self-appropriation led him instead to the discovery that his own knowing—and human knowing in general—situates seeing and touching within a larger, dynamic "cognitional structure" in combination with other operations and activities of consciousness. That structure consists of:

- Seeing and touching, but also other sensations, imagination, and memory;
- followed then by questions (what, why, and how) about those experiences;
- which lead to insights;
- succeeded by questions about whether the insights are correct;
- which lead to other activities ("reflecting") involved in checking one's past experiences and seeking new ones to fill out evidence;
- reflecting also involves further questions seeking further insights that will modify, correct, or replace one's initial insights;
- finally judgments that affirm some sets of insights as really, factually correct, but only insofar as this process of reflecting has reached its natural limit of answering all the pertinent questions squarely.
- The various activities in this structure fall into one of three "levels of consciousness"—experiential, intellectual, reasonable (or critical);
- questions move thinking back and forth among these levels in a self-correcting pattern as insights are refined and modified toward eventual correctness.[6]
- Objectivity in knowing is "intrinsic" to this structured process, because questions set the criteria that any claim to knowledge must satisfy.[7]

The questions in this dynamic structure connect the activities to one another and thereby make it a cyclical, dynamic, and self-correcting structure that repeats with each occurrence of new questions. Because questions play the leading role in this structure, the structure's dynamism is both self-correcting and self-transcending. It is self-transcending because questions draw thinkers beyond their sensations toward understandings of those sensations and beyond the merely possible understandings toward correct and true understandings. It is self-correcting because the intended end point of each cycle of this dynamic structure is a judgment that genuinely answers all the relevant questions and thereby affirms some set of insights as true of reality. The dynamic structure transcends beyond what a person merely thinks into authentic knowledge of reality itself.

Human beings spontaneously use the operations of consciousness in this structure as they make sense of what they have received from their traditions and heritages, but they are seldom fully aware of this structure in all its details. Lonergan proposed self-appropriation as the way to assist people in becoming explicitly knowledgeable of the structure of activities they spontaneously use and the important implications this has about reality and objectivity. He argued that this heightened self-understanding lays the ground for a more effective evaluation of what is received from the past and what should be done in the future.

Initially Lonergan thought this cognitional structure also underpinned the ways that people arrive at judgments about matters of value and ethics. A few years after the publication of *Insight*, however, he realized that questions about values bring to life a fourth and distinct level of consciousness and an expansion in the structure of cognition. He called this the fourth level of deliberation and responsibility. I fleshed out his suggestions in *The Ethics of Discernment* into what I call the structure of ethical intentionality.[8] Briefly, the structure of ethical intentionality begins with the use of cognitional structure in order to know one's situation accurately. It then goes on to add the further structured activities of the fourth level:

- Questions about what one could do;
- followed by insights about possible courses of action one could take;
- next to questions about whether those courses of action should be undertaken and questions about what values (or disvalues) such actions would bring about;
- then to activities of deliberation and value reflection to answer satisfactorily all those questions about the values of what one should do;

- finally to judgments of value, decisions, and actions that make the chosen values realities.

- Throughout the totality of this process "feelings as intentional responses to values"[9] play an indispensable role, providing the most elemental apprehensions of values. While feelings alone do not yield objective knowledge of values, such knowledge cannot be attained without their inputs.

These are the operations in the structure of ethical intentionality, the fourth level of consciousness, that people spontaneously utilize as they evaluate the ideas and practices they receive from their predecessors and as they think about how they should respond.

Self-appropriation facilitates explicit knowledge and deliberate commitment to one's own structures of cognition and ethical intentionality. It begins with exercises in heightening one's self-awareness of the various activities of consciousness that go on in spontaneous performances of these structures. It then endeavors to understand and know correctly what is brought to light in this heightened awareness. It then invites one to affirm the inherent values of these structures and to commit oneself to living in fidelity and consistently to the questions and procedures of these structures. Explicit self-appropriation of these structures has further profound implications for how one conceives of what is real, good, true, objective, ethical, and right.

3. From Structures of Consciousness to Functional Specialties

Lonergan's method of eight integrated functional specialties is a sophisticated, self-reflective, and deliberate method based upon the structures of cognition and ethical intentionality that ordinary people and scholars already use spontaneously in evaluating their traditions and deciding what actions to take moving forward. Lonergan was inspired to organize his functional specialties by his identification of the four levels of consciousness within the structures of cognition and ethical intentionality: four levels in evaluating inheritances from the past, and another four levels in deciding what actions to take toward the future. Hence, Research, Interpretation, History, and Dialectic are intended as highly specialized methods to produce enhanced results in the work of evaluating the past that correspond to the experiential, intellectual, reasonable (critical), and responsible levels respectively. Likewise, the specialty of Foundations corresponds to the level of responsibility where fundamental commitments are made. Policy hammers out judgments and directives that are determined by Foundational commitments. Planning seeks creative insights that will organize the means needed to realize policies. And Implementation brings about the experiential consequences of plans put into action.

Organizing the functional specialties according to the pattern of structures of consciousness does not mean that any one of the specialties is limited to the activities on its corresponding level of consciousness alone. The entirety of cognitional and ethical intentionality structures is used in each of the functional specialties. For example, researchers do more than experience expressions of past ideas. They have to have insights and make judgments about whether the expressions came from the alleged sources and the times ascribed. Still, the primary objective of Research is to provide reliable data to the later functional specialties.

Fuller details on each of these functional specialties will follow momentarily. Let me observe for the moment, however, that Lonergan's use of the levels of consciousness does make a certain amount of sense as a way of organizing the functional specialties. In the ordinary, spontaneous exercise of the structures of cognition and ethical intentionality, each of the levels of consciousness produces contents (experiences, insights, judgments of fact, judgments of value, and decisions) that are passed along as material to the next level. Questions pass experiences along to intelligence in search of insights. Further questions pass insights along to solicit judgments about their factual correctness. Knowledge of factual situations elicits questions about what courses of action and values should be chosen and enacted.

We now turn to each of the functional specialties in detail.

(1) The specialty of Research incorporates the techniques that have been developed by scholars to determine the authenticity of the data on the expressions from people in the past. Research seeks out and eliminates anything that might have corrupted the physical texts, drawings, recordings, and so forth, during its transmission, such as physical damage, alteration (whether unintentional or fraudulent) in copying, or embellishment of what was originally spoken, written, or drawn.

(2) Interpretation takes up the refined data on expressions passed along from Research and seeks understanding in response to the question

"What does this mean?" It uses the many exegetical and hermeneutical techniques that have been developed over the past two centuries—especially those pertaining to interpreting meanings in their original contexts—in order to offer the best possible understandings of what was spoken or written in the past.[10]

(3) Historians utilize methods from both of the first two functional specialties as regular parts of their work, but History in Lonergan's technical sense specializes in the work of narrative. History as a functional specialty draws upon scholarly interpretations of meanings in order to discover the critically grounded narratives that explain how these meanings are related to one another over the course of time. History is concerned to know the facts as they really happened, but it is concerned with more than that. History is primarily the work of finding the narrative (or complementary narratives) that correctly understand how the events are meaningfully connected to one another across time.

Some professional historians employ all three of the first three functional specialties in their work, although without the same degree of reflective self-awareness fostered by Lonergan's framework of eight functional specialties. Many, however, do one kind of work while relying upon the results of others. (Joseph DesJardins, for example, did not do the work of authenticating the texts he used in developing his narrative of the history of environmental ethics relied upon in part 1 of this book.)

Lonergan especially esteemed the ways that modern historians play crucial roles in critical knowledge and evaluation of what the past has bequeathed to us, and he therefore gave the functional specialty of History a prominent place within the larger integrated framework of his eight functional specialties. One of Lonergan's own contributions was to show how the structures of cognition and ethical intentionality operate in the work of historians, and to show how this expanded self-awareness can enhance the works of historians. He showed, for example, that a full self-appropriation the structures of consciousness as employed by historians does not imply historicism, contrary to what many have thought, historians included.[11] He did something similar for the other seven functional specialties as well.

(4) Dialectic addresses conflicts in the historical movements studied as well as disagreements among the scholars about the meanings of those movements and events. Some conflicts are due to cultural differences. Others are due to one or more of what Lonergan called "biases": the psychological, individual, group, and general biases. In his technical

sense of the term, a bias is a major interference with the self-correcting dynamics of ethical intentionality. Each of the four biases employs a distinctive mechanism of interference.[12]

Psychological bias (which Lonergan calls "dramatic bias") blocks memories or imaginings that would lead to uncomfortable or undesirable insights or feelings. Psychological bias can even cause people to overlook or "tune out" things they don't wish to see or hear, even things that are obvious to everyone else. The basic strategy of psychological bias is to fill up one's consciousness with images or memories that will lead away from unwanted insights or feelings. Almost everyone has some degree of psychological bias that makes them unaware of certain aspects of their environment, but psychological biases can also grow into severe mental illnesses.

Individual bias is known more commonly as selfishness or egoism. Individual bias is often spoken of as caring only about one's own self. This means pursuing insights and courses of action that will benefit one's own immediate desires and interests, and allay one's own narrow fears without regard to those of anyone else. This is really not caring only about oneself in the fullest sense, but caring only about a self defined narrowly by one's own desires and fears alone.

Group biases grow out of exclusive affection for people like oneself while lacking in sympathy or empathy for "others" who are different. Group biases include the "isms" such as racism, sexism, classism, homophobia, ethnocentrism, nationalism, anthropocentrism, and so on.

General bias originates in the notable success that common sense has in practical matters. However common sense tends "to rationalize its limitations by engendering a conviction that other forms of human knowledge are useless or doubtfully valid." As Lonergan puts it, general bias "tends to set common sense against science and philosophy." Because of general bias, common sense runs the risk of extending "its legitimate concern for the concrete and the immediately practical into disregard of larger issues and indifference to long-term results."[13] Dialectic as method traces the tensions and hostilities in many human affairs back to these sources of bias. In doing so, it shows that such conflicts cannot be easily resolved by rational arguments or factual evidence. The roots of the biases must be first identified, and often they can only be overcome by intellectual, moral, or even religious conversions.

In addition to the conflicts that arise out of biases, still other conflicts arise among people who differ in terms of what Lonergan called "differentiations of consciousness." These are broad ways in which the fundamental structures of cognition and ethical intentionality have been developed over the long course of history to operate within different realms of meaning, including the religious, artistic, theoretical/scientific, scholarly, and interiorly differentiated realms. Lonergan considers common sense "undifferentiated" because it engages partially in each realm without full refinement in any of them, and it tends to confuse meanings from one realm with meanings from another. Here conflicts arise because "undifferentiated consciousness . . . is at home in its local variety of common sense but finds any message from the worlds of theory, of interiority, of transcendence both alien and incomprehensible."[14] (Chapters 14 and 15 identify undifferentiated consciousness as a source of conflict between science and common sense.)

The most fundamental conflicts arise out of the basic commitments people have made, tacitly or explicitly, regarding knowledge, reason, reality, goodness, truth, objectivity, and ethics. Dialectic endeavors to identify the sources of these conflicts and make those sources explicit, for their power is amplified when they remain hidden. In Dialectic these unacknowledged commitments are brought into the light successfully if its practitioners themselves have appropriated how they are using or neglecting their own structure of ethical intentionality in making evaluative judgments. Dialectic as such takes no stand on which of the opposed stances regarding the fundamental questions is the correct one; rather, it exposes those oppositions as clearly as possible. Chapter 8 offers a more complete and more concrete illustration of how the Dialectical method can be applied to the history of environmental ethics.

The second set of functional specialties also reflects the ways each level of the structures of cognition and ethical intentionality passes something along to the next, but in an asymmetrical fashion. In this case the correspondence with the levels of consciousness is reversed. People make judgments about what is true in fact and in value (Policies) on the basis of their fundamental commitments (Foundations). They formulate plans of action (Planning) on the basis of their policy judgments. And they undertake concrete, experienceable actions (Implementation) on the basis of the plans to which they have committed.

(5) Foundations takes on the task of making explicit, not someone else's fundamental commitments, but one's own. It draws upon Lonergan's realization that there are fundamental turning points, conversions, through which those fundamental commitments undergo dramatic

change. This is illustrated poignantly in chapters 3 and 4 in the conversions of Albert Schweitzer, John Muir, and Aldo Leopold toward a value-wholeness that transcended their previous horizons of value. Lonergan identified intellectual, moral, and religious conversion as the most fundamental of such turning points. The functional specialty of Foundations, therefore, endeavors to express as clearly as possible the nature and implications of these conversions in order to "make conversion a topic."[15] If Foundations is to be methodical, it cannot preempt what is or is not a conversion in matters of reality, truth, knowledge, or the good. Such claims must be the outcomes of complex dialogues. Foundations is a heuristic methodical specialty to facilitate such dialogues, not to dictate answers to the relevant questions.[16] At the same time, it must take seriously that people do have such fundamental commitments, and that the process of self-appropriation raises new kinds of awareness about the strengths, weaknesses, and implications of such commitments.

(6) In the field of theology, the functional specialty Doctrines uses the clarification of fundamental commitments brought to light in Foundations in order to select as one's own some specific judgments from one's heritage as they have been criticized and refined through Research, Interpretation, History, and Dialectic.

(7) Systematics seeks the best possible understanding of the doctrines chosen, especially when there is need to reconcile apparent conflicts among the doctrines one has endorsed, and apparent conflicts between those doctrinal commitments and knowledge in other areas, such as natural and social science, historical studies, and fields of common sense.

(8) Communications develops ways to express these systematized understandings of doctrines in nonsystematic ways to people in a wide variety of cultures and walks of life. Communications is the culmination of the eight functional specialties in theology because the fundamental objective of Christian theology is to preach the good news to all nations. All of Christian theology is oriented toward this ultimate goal.

Lonergan realized, however, that this method for evaluating the past and arriving at the right courses of action for the future need not be limited to theology; it also opens out into the fields of humanities, social sciences, ethics, and public policy. This is possible because the method of functional specialties does not presuppose the doctrines of any particular religious tradition. Lonergan writes, "Indeed, the functional specialties of Research, Interpretation, and History can be applied to the data of any sphere of scholarly human studies."[17] Likewise, the last three functional specialties can be conceived in ways that make them also relevant to all areas of human endeavors, with appropriate adjustments. "Corresponding to doctrines, systematics, and communications in theological method, integrated studies would distinguish policy making, planning, and the execution [or implementation] of the plans."[18] From this expansion, it follows:

(6') While groups make policies all the time, Policy as a functional specialty adds a self-reflective methodical dimension to this process. The methodical specialty of Policy draws upon prior claims about realities and values that have been refined and criticized by the first three functional specialties along with the sources of conflict exposed by the fourth, Dialectic. It also draws upon the results of Foundations to discriminate among the refined sources and to arrive at strategic propositions (policies) that will guide future actions. Examples of policies for environmental action are the UN Sustainable Development Goals discussed in chapter 19. While the UN did not use Lonergan's method of Policy to arrive at these goals, in that chapter I indicate how the method can further enhance this process.

(7') Planning builds upon the strategic judgments arrived at in Policy by adding the large number of coordinated insights into the means that will be needed to bring the policies into realities. Chapter 19 also offers examples of how planning builds upon policies in the field of environmental ethics in the instances of the targets and indicators being used to make the UN Sustainable Development Goals more concrete.

(8') Implementation is most noticeable in the many and diverse actions taken to realize plans. But the real outcome of Implementation is not the actions themselves, but consequences of those actions that change the world and the experiences people have of that world.

The wider, nontheological scope of Lonergan's method just outlined is important, since most of the contemporary philosophical debates about ethical matters in general, and environmental ethics in particular, are pursued almost exclusively by means of secular approaches. This does not at all mean that theological contributions to environmental ethics are irrelevant. Quite to the contrary, chapter 18 shows how Pope Francis' theological encyclical *Laudato si'* has made substantial contributions to the history of environmental ethics. Still, the broadening of the method does make it possible to treat methodically the many contributions and options that do not rely upon any specific theological tradition.

Pivotal between the first three functional specialties (Research, Interpretation, History) and the last three (Policy, Planning, Implementation) are the two intermediate specialties of Dialectics and Foundations. They perform the crucial roles of connecting all eight into an integrated, methodical whole. These are of the utmost importance not only in the field of theology but also in all areas of human studies, including environmental ethics. As Lonergan says:

> In historical and empirical human studies scholars and scientists do not always agree. Here too, then, there is a place for dialectic that assembles differences, classifies them, goes to their roots, and pushes them to extremes by developing alleged positions while reversing alleged counterpositions. [Likewise, foundations], which objectify the horizon implicit in religious, moral, and intellectual conversion, may now be invoked to decide which really are the positions and which really are the counterpositions. In this fashion any ideological intrusion into scholarly or scientific human studies is filtered out. The notion of dialectic, however, may play a further role. It can be an instrument for the analysis of social process and the social situation. The social historian will ferret out instances in which ideology has been at work. The social scientist will trace its effects in the social situation. The policy maker will devise procedures both for the liquidation of the evil effects and for remedying the alienation that is their source.[19]

The meanings of Lonergan's technical terms, position and counterposition, were explained at length in chapter 8. But his references to the possible presence of positions and counterpositions in the preceding quotation does not mean that Lonergan presumed to teach historians how they should do their work. He pointed out that historians already grapple with counterpositions in their work as they "ferret out" ideologies. Indeed, he based his own reflections on the first three functional specialties (and especially History) on the extensive and valuable work already conducted by historians and historiographers long before he formulated his eightfold method. His great contribution, rather, was to show how historical studies can be methodically situated within a wider heuristic framework by also drawing explicit attention to claims that are performative contradictions of the structures of cognition and ethical intentionality.

Historical studies reveal the plurality and development of ethical principles, for example,[20] but historians qua historians do not take an explicit stance on those principles. To move *methodically* from the plurality and developments of ethical principles toward taking a stance requires the further contributions of Dialectic and Foundations. Only once those contributions have been added to what historians have provided can the functional specialty of Policy proceed in a methodical fashion. As a methodical specialty, Policy looks to the work of historians, as criticized and refined by Dialectic, for the ethical precepts that have arisen historically. Drawing on Foundations, Policy scrutinizes such principles to determine which ones are compatible and which incompatible with the "positions." Formulations of what count as "positions" are regularly refined as Foundations endeavors to better understand the authentic, converted forms of human subjectivity.[21] Planning seeks insights into fundamental strategies that are needed to realize policies. Finally, Implementation would necessarily require the coordinated actions of many people, especially in the realm of environmental ethics. These many people need to be persuaded of the value of the actions they are being asked to perform. This means that Implementation must include great efforts at persuasion and communication (the final functional specialty for theology) but will also include efforts that produce the changes needed in human institutions as well as the physical environment in order to realize plans and policies. All three functional specialties—Policy, Planning, and Implementation—require the collaboration of large numbers of people. They require their extensive efforts in figuring out and contributing vast numbers of insights and judgments of facts and values to the ongoing process of collectively addressing the needs of the present going into the future.

In part 1, I relied upon DesJardins and Cittadino to provide histories of the dialectical development of environmental thought as it pivots around momentous conversions to holism. In Part 2 I drew upon the works of a number of historians to show how the crisis in climate change ethics pivots around the meanings of certainty, probability, trust, and belief in science. In both parts 1 and 2, I endeavored to show what Dialectic offers in the method of ethics in general and environmental ethics in particular.

Lonergan placed the eight functional specialties in a particular order, from Research through Implementation, but the direction need not be unidirectional. For example, researchers, interpreters, historians, and dialecticians will also be better equipped in their work insofar as they are increasingly able to operate with self-reflective awareness of Foundational positions rooted in conversions.

Most of this book is focused on the functional specialties of History, Dialectic, and Foundations, but illustrations of Policy, Planning, and Implementation are addressed in chapters 18 and 19. My own application of Dialectic to the works of the historians relies heavily on how I have understood Lonergan's account of Foundations, so in the next section I offer a brief summary of some of the key Foundational ideas I use at various places in this book.

4. Foundations for Environmental Ethics

For Lonergan, the real foundations of theology—and, by extension, for environmental ethics—are converted, authentic subjects.[22] It is the job of the functional specialty Foundations to formulate the basic "categories" that are implicit in these conversions. Although much needs to be said about theological dimensions of environmental ethics, here I will limit myself to what can be said about the philosophical Foundations of environmental ethics on the basis of intellectual and moral conversions.[23]

There are three central Foundational categories that are relevant to environmental ethics: (1) Lonergan's own account of cognitional structure; (2) what I have called the "structure of ethical intentionality"; and (3) what Lonergan called "the ontology of the good," which I call "the emerging good" in chapter 7. Cognitional structure and the structure of ethical intentionality are outlined in section 2 of this appendix and also in chapter 1, so I will not expand further on those here, but they do have further Foundational implications.

Foundations takes the stance that human beings spontaneously use the operations of consciousness in cognitional structure as they endeavor to understand and know the world of their experiences. Yet people are seldom fully aware of this structure in all its details or of all the implications about reality and objectivity that are tacit in their reliance on this dynamic structure. Affirming cognitional structure as the "position" about human knowing has the implication that reality is intelligible. This is because each true, objective judgment of fact is always an affirmation that the intelligible content of some insight "is really so." Reality known in this fashion is intelligible rather than brute matters of material facts. The intelligibility of reality has further implications for what the empirical sciences know about reality. The methods used by modern sciences imply that the natural world has the general intelligible structure of emergent probability as explained in chapter 7.

The Foundational claim that people use the structure of ethical intentionality in their deliberating and deciding also has Foundational implications. Just as most people assume reality is limited to visible and tangible materiality, they likewise tend to think that the good boils down to the bodily pleasures that meet their immediate desires and needs. Again, just as self-appropriation leads to a radical reorientation in one's conception of reality as intelligible, it likewise leads to a radical reorientation from the good-as-pleasant to the vast totality of the good that is sought by questions for value and deliberation. The structure of ethical intentionality implies that the good is a complex, interconnected, emerging set of values that I have called "the emerging good" with its accompanying scale of values. That scale of values has its primordial manifestation in the feelings of value preference that emerge spontaneously in human consciousness, although feelings are also subject to radical distortions and inversions. I have outlined this expanded sense of the wholeness of the good in chapter 7.

These reorientations regarding reality and the good will be disorienting, as was the case for the prisoner who had to be "turned around" (περιαγωγή) and dragged out of the cave in Plato's *Republic*.[24] Lonergan therefore came to speak of accepting the implications of one's self-appropriation of the structures of cognition as "intellectual conversion" and of ethical intentionality as "moral conversions." These conversions are not irrational abandonments of the norms of rationality, but rather conversions to exactly what those norms truly are. Rationality is taking every question seriously, following its lead through various possibilities and alternatives until all pertinent questions of fact and value have been answered. Rationality reaches its culmination in objective judgments of fact and value, and authentically ethical decisions.[25] When Lonergan speaks of "conversion," he means abandoning all the self-serving and self-deceiving modes of discourse that pass for factual and ethical rationality and their erroneous implications, adhering instead to the lead of ever further questions.

5. Conclusion

Much of the history of environmental ethics is the history of attempts to answer questions about the proper place of human beings and their actions in relationship to the rest of nature. It is also a history of the criticisms of the destructive consequences human actions have had on nature when such questions were not properly considered or answered. Asking questions about responsibility for the environment and being obligated to act only on the basis of good answers to those questions—this is not something any other creatures do. Hence the special value of human beings has to do with their special responsibilities toward nature and each other, and not at all with the special entitlements or privileges they might claim for themselves to dominate the rest of nature. Human beings have the special value of being stewards of the continuation of the emerging good. This book has endeavored to show how Lonergan's method of eight functional specialties can enhance and guide this office of stewardship.

Notes

Introduction: Environmental Ethics as Historical

1. The document coming out of the 2015 Paris Conference of the Parties (COP21) was adopted as a treaty by many nations. In the United States, however, treaties must be ratified by two-thirds of the Senate. Since the Obama administration did not believe that two-thirds of senators would vote to approve the Paris document, it was instead accepted as a presidential "agreement." (The "parties" in "Conference of the Parties" refers to nations that adopted the United Nations Framework Convention on Climate Change (UNFCCC) at the United Nations Conference on Environment and Development (UNCED), also known as the Earth Summit in Rio de Janeiro, in 1992. The COP is held annually (except for 2020 because of the COVID pandemic) to report progress and update commitments growing out of UNFCCC.

2. Spencer Weart, *The Discovery of Global Warming*, revised and expanded edition (Cambridge, MA: Harvard University Press, 2008), viii. Cited hereafter as Weart.

3. Patrick H. Byrne, *The Ethics of Discernment: Lonergan's Foundations for Ethics* (Toronto: University of Toronto Press, 2016), 413–48. Cited hereafter as *EOD*.

4. Alexis de Tocqueville noted this cultural prejudice in the mid-nineteenth century. Speaking of Americans who see themselves as "self-sufficient," Tocqueville writes, "These owe nothing to anyone, they expect so to speak nothing from anyone; they are in the habit of always considering themselves in isolation, and they willingly fancy that their whole destiny is in their hands." Alexis de Tocqueville, *Democracy in America*, trans. Harvey C. Mansfield and Delba Winthrop (Chicago: University of Chicago Press, 2000), 484. Hans-Georg Gadamer expands the range of this claim, identifying a prejudice as our inheritance from the Enlightenment, and arguing that "the fundamental prejudice of the enlightenment is the prejudice against prejudice itself, which deprives

tradition of its power." Hans-Georg Gadamer, *Truth and Method* (New York: Seabury Press, 1975), 239–40.

5. See Tad Dunne, *Doing Better: The Next Revolution in Ethics* (Milwaukee: Marquette University Press, 2010).

6. I will capitalize terms referring specifically to the eight functional specialties as Lonergan formulated them.

Chapter 1: From Scientific Facts to Ethical Values

1. See the appendix to this volume, "Method in Environmental Ethics."

2. Eugene Cittadino, "Ecology and American Social Thought," in *Religion and the New Ecology: Environmental Responsibility in a World in Flux* (Notre Dame, IN: University of Notre Dame Press, 2006), 73–115. Cited hereafter as EAST. Joseph R. DesJardins, *Environmental Ethics: An Introduction to Environmental Philosophy*, fifth edition (Boston, MA: Wadsworth, 2013), cited hereafter as *EE*. Although DesJardins' book is an introductory textbook in environmental ethics, it is organized into a well-researched narrative of the movement of the history of environmental thought that covers the same time period as does Cittadino. While there are numerous historical studies of environmental ethics, I have chosen DesJardins' and Cittadino's works because they are the most comprehensive and most recent studies of the engagement of environmental ethics with the evolving environmental sciences.

3. *EE*, 7.

4. *EE*, 5–6.

5. Rachel Carson, *Silent Spring*, fiftieth anniversary ed. (Boston: Mariner Books, Houghton Mifflin Harcourt, 2001). Cited hereafter as Carson.

6. *EE*, 9–11.

7. A full implementation of Dialectic in the field of environmental ethics would require consideration of other histories of environmental science and ethics. In an extended footnote, Cittadino briefly surveys several additional histories of ecological science and environmentalism. He mentions the various commitments that underlie the accounts and countercritiques among these studies. See EAST, 106–07, footnote 5.

8. *EE*, 31.

9. David Hume, *A Treatise of Human Nature* (London: Clarendon Press, 1975), 457. Cited hereafter as Hume. In all citations from Hume's *Treatise*, I have replaced his contractions with their modern-day equivalents.

10. Hume, 1.

11. Hume, 1.

12. Bernard Lonergan *Insight: A Study of Human Understanding*, Collected Works of Bernard Lonergan, vol. 3, edited by Frederick E. Crowe and Robert

M. Doran (Toronto: University of Toronto Press, 1992), 111. Cited hereafter as *Insight*.

13. Hume, xv. For my own contrary view, see *EOD*, 98–14 and the section "An Alternative Account of Scientific and Ethical Reasoning" in this chapter. My approach answers Hume's criticism because it does not confine ethical reasoning to logical deduction. Rather, it expands reasoning outward into the various self-correcting processes of question-and-answer. This provides a more basic foundation for ethics than either that of Hume or of other prevailing ethical theories that rely upon rigorous logical deduction as the paradigm of "reason."

14. Hume, 469–70.

15. Hume, 468, emphasis added.

16. Hume, 457.

17. Hume, 415.

18. Hume, 459.

19. *Insight*, 389.

20. Bernard Lonergan, *Method in Theology* (New York: Herder and Herder, 1972), 238. Cited hereafter as *MT*. See also *Insight*, 440–41 and 449–50. This section, therefore, is an application of the functional specialty of Dialectic, insofar as it identifies the "counterpositions" about reasoning in Hume's work and proposes the alternative "position" about reasoning to judgments about facts and values. See chapter 8.

21. *Insight*, 44.

22. *Insight*, 197; see also 441, 449–50.

23. *Insight*, 309.

24. *Insight*, 214–20 and 244–57.

25. "Such a procedure, clearly, is logical, if by 'logical' you mean 'intelligent and reasonable.' With equal clearness, such a procedure is not logical, if by 'logical' you mean conformity to a set of general rules valid in every instance of a defined range; for no set of general rules can keep pace with the resourcefulness of intelligence in its adaptations to the possibilities and exigencies of concrete tasks of self-communication." *Insight*, 201.

26. *EOD*, 95–114; see also my "Desiring and Practical Reason: MacIntyre and Lonergan," *International Philosophical Quarterly* 60, no. 1, issue 237 (March 2020): 75–96.

27. See *EOD*, 207–40.

Chapter 2: Environmental Ethics in the Utilitarian Mode

1. *EE*, 23–43.

2. *EE*, 27–32 and 37–40.

3. *EE*, 36. See also: "Utilitarian reasoning is especially influential in the areas of economics, public policy, and government regulations, and this means that it has also played a significant role in environmental policy" (33).

4. As quoted in *EE*, 52.

5. For the difference between preservationism and conservationism, see chapter 3.

6. *EE*, 53.

7. In terms of Lonergan's method of functional specialties, we would say that the ethical framework of utilitarianism functioned as a Foundational category about the good, and that fire suppression derived from that foundation as a Policy. See the appendix, "A Method for Environmental Ethics."

8. The US Forest Service policy of fire suppression has since been replaced in the light of more recent forest science research about the long-term role of fires in forest ecosystems.

9. Jeremy Bentham, A *Fragment on Government: An Introduction to the Principles of Morals and Legislation*, preface, in *The Collected Works of Jeremy Bentham*, electronic edition, 393, http://library.nlx.com.eu1.proxy.openathens.net/xtf/view?docId=bentham/bentham.01.xml;chunk.id=div.comment.commentaries.21;toc.id=div.comment.commentaries.21;brand=default, accessed December 5, 2021.

10. John Stuart Mill, *Utilitarianism* (Kitchener, ON: Batoche Books, 2001), 11.

11. *EE*, 34–35.

12. *EE*, 34.

13. Using pleasure and pain as the indices for objective ethical judgments and decisions amounts to what Lonergan called a "counterposition." For the more authentic "position" regarding objectivity in ethics, see *EOD*, chapter 8.

14. As quoted in *EE*, 56.

15. *EE*, 56–57.

16. Lonergan's methodological specialty of Dialectic was designed to address conflicts such as those posed between the policies advocated by Pinchot and O'Toole. But theirs is not at what Lonergan regarded as the deepest level of conflict—between "positions" and "counterpositions" as Lonergan identified them. For further details, see chapter 8.

17. *EE*, 36.

18. *EE*, 81.

19. *EE*, 82.

20. *EE*, 255, emphasis added. DesJardins does not discuss "consequentialism" or its role in the history of environmental ethics. Consequentialism is a somewhat later revision of utilitarianism. Instead of pleasures and pains, consequentialism seeks to maximize other kinds of goods or values. Within consequentialism there is much debate about what should count as the goods or values to be added into the calculus. In *EOD* I have presented an alternative way of conceiving of the wholeness of the good, including values and a ground for objectively evaluat-

ing differing values. See *EOD*, chapters 9, 13, and 14. DesJardins does briefly consider one alternative to the problem of responsibilities to future generations offered by theories of rights in the deontological tradition. But he shows that deontology runs against a different problem than utilitarianism in claiming that future generations have rights to a decent environment: future generations do not yet exist, so they cannot be said to *have* anything, including rights. While several deontologists have attempted to address this problem, DesJardins again does not think their attempts have been completely satisfactory.

21. What I have called the "ethics of discernment" was developed by use of the functional specialty of Foundations. See *EOD*, 443–47.

22. See *EE*, 99.

23. DesJardins also criticizes Kant for this same failure of anthropocentrism in his approach to ethics.

24. *EE*, 100.

25. *EE*, 111.

26. As quoted in *EE*, 107. Although DesJardins says Feinberg's approach is more closely associated with a deontological/rights rather than a utilitarian approach to ethics, it is hard to see the continuity with the Kantian deontological tradition where the capacity of human *reason* to grasp and show respect for duty, law, and the categorical imperative are so central. Surely Kant would have regarded "wishes, desires, hopes" as inclinations having no moral worth. Contrary to DesJardins' categorization, therefore, Feinberg should more properly be understood as a utilitarian extensionist.

27. As quoted in *EE*, 114. As with Feinberg, it is difficult to see how Regan is really within the deontological tradition, since "desires, perception, memory, and a sense of the future . . . emotional life feelings of pleasure and pain, preference and interests," fall more in line with what Kant called inclinations. Regan also, therefore, is more accurately classified as a utilitarian extensionist.

28. *EE*, 112.

29. *EE*, 114.

30. As quoted in *EE*, 118.

31. *EE*, 119, emphasis added. On the shift from satisfaction toward value, see *MT*, 240, and *EOD*, 227–32.

32. *EE*, 160–61, 180.

33. *EE*, 119.

Chapter 3: Conversion to the Value of Life as a Whole

1. A third person who also wrote a vivid description of his holistic conversion experience, Aldo Leopold (1887–1948), comes a bit later in DesJardins' historical narrative. Hence discussion of his work will be deferred to chapter 4.

2. *EE*, 133.

3. *EE*, 133.

4. *EE*, 133.

5. *EE*, 134.

6. Concerning the importance of an "explanatory" approach to ethics and metaphysics, see *Insight*, 528–33.

7. *EE*, 138.

8. *Insight*, 61; see also 103.

9. Notice, however, that Taylor overlooks the problem of how one can reason from biological knowledge of the facts of specific life cycles to knowledge of these ethical principles or even to knowledge of the values of what "benefits or harms them, what environmental changes are to their advantage or disadvantage." This issue is addressed in chapter 1.

10. *EE*, 143. See 136–42 for his full discussion of Taylor's positions.

11. From Denis C. Williams, *God's Wilds: John Muir's Vision of Nature* (College Station: Texas A&M University Press, 2002), as quoted at https://en.wikipedia.org/wiki/John_Muir#cite_note-Williams-59.

12. John Muir, *My First Summer in the Sierra*, 5, June 6, emphasis added, http://www.yosemite.ca.us/john_muir_writings/my_first_summer_in_the_sierra/my_first_summer_in_the_sierra.pdf.

13. *EE*, 52.

14. *EE*, 151–52.

15. John Muir, "The Hetch Hetchy Valley," *Sierra Club Bulletin* 6, no. 4 (January 1908): 211–20.

16. Given the eventual importance of Muir to environmentalism, DesJardins' treatment of him is uncommonly brief. I have therefore added a few supplemental comments to his own narrative. However, since it is my intention to show how Lonergan's method builds upon the work of historians, I did not venture far beyond DesJardins' own narrative.

17. *EE*, 153, citing the Wilderness Act of 1964. He presents other versions of the same conception.

18. *EE*, 154; see also Richard Slotkin, *Regeneration through Violence: The Mythology of the American Frontier, 1600–1860* (Norman: University of Oklahoma Press, 1973).

19. *EE*, 154–55.

20. *EE*, 157

21. *EE*, 157.

22. EAST, 80, and *EE*, 156.

23. EAST, 75.

24. *EE*, 158.

25. *EE*, 154.

26. *EE*, 160; see also EAST, 98.

Chapter 4: The Science of Ecology and Conversion to Environment as a Whole

1. EAST, 74.
2. *EE*, 164; see also EAST, 76–77.
3. EAST, 78–79.
4. *EE*, 165.
5. EAST, 82.
6. *EE*, 165.
7. EAST, 81. These scientific ideas would later become significant for Aldo Leopold's "land ethic."
8. *EE*, 166.
9. EAST, 83.
10. *EE*, 167.
11. *EE*, 168; see also EAST, 83–85.
12. EAST, 83.
13. *EE*, 170.
14. *EE*, 170–71.
15. *EE*, 179. DesJardins writes that Leopold's 1933 book *Game Management* became the classic text in the field.
16. Aldo Leopold, *Sand County Almanac and Sketches Here and There: Special Commemorative Edition* (New York: Oxford University Press, 1989; originally published in 1948), 129. Cited hereafter as Leopold.
17. Leopold, 129–30.
18. In his article "Wolves, Wisconsin, and Aldo Leopold," Christian Diehm has argued that Leopold's conversion was not all at once on the occasion of the killing of this particular wolf. Rather, "Leopold's shift away from the belief that predators were pests with no redeeming value whatsoever—[is] something that did happen, but not suddenly, and not as early as 1909." Quoted from the blog post, https://www.humansandnature.org/wolves-wisconsin-and-aldo-leopold, viewed April 19, 2019. Originally published in *Minding Nature* 6, no. 2 (Spring 2013): 44–48.
19. *EE*, 180.
20. *EE*, 180.
21. See *EOD*, 136–68.
22. *EE*, 181.
23. *EE*, 181, 182.
24. Cittadino narrates how the residual organic model has endured beyond Leopold in EAST, 91–95, 99.
25. EAST, 86–87.
26. *EE*, 161.

Chapter 5: Deep Ecology and Ecofeminism

1. See *EOD*, 144–47, on feelings as intentional responses to judgments of fact.
2. *EE*, 206.
3. *EE*, 207.
4. *EE*, 208–12.
5. *EE*, 210.
6. *EE*, 210, 212, emphasis added.
7. *EE*, 212.
8. *EE*, 211.
9. *EE*, 212.
10. *EE*, 209
11. A performative contradiction is a contradiction between what one claims and the operations one uses to arrive at that claim. Lonergan coined the technical term *counterposition* to denote performative contradictions. For details see chapter 8. In chapter 1 we pointed out another instance of performative contradiction in Hume.
12. *EE*, 215.
13. *EE*, 218, 219.
14. *EE*, 220.
15. *EE*, 220.
16. *EE*, 225.
17. *EE*, 221.
18. *EE*, 225.
19. *EE*, 222.
20. *EE*, 222, quoting from Val Plumwood, "Current Trends in Ecofeminism," *The Ecologist* 22, no. 1 (January–February 1992): 10.
21. *EE*, 223.
22. *EE*, 223–24.
23. *EE*, 225.
24. *EE*, 224–26.
25. *EE*, 228.

Chapter 6: Environmental Justice, Environmental Racism, and a Pragmatist Compromise

1. *EE*, 234.
2. *EE*, 235.
3. John Locke, *Second Treatise of Government*, chapter 2, §§4–6.
4. *EE*, 236.

5. *EE*, 235.

6. *EE*, 234.

7. *EE*, 239. Presumably Rawls' theory would support this because the potential burdens from this proposed action would be outweighed by the potential benefits, and that both should be distributed equally, and showing that this is the case must be done with precision.

8. Michael J. Sandel, *Liberalism and the Limits of Justice* (New York: Cambridge University Press, 1998).

9. *EE*, 240.

10. *EE*, 240.

11. *EE*, 240.

12. *EE*, 241.

13. *EE*, 241.

14. *EE*, 242.

15. *EE*, 242.

16. *EE*, 242.

17. For a treatment of what would be sufficient reasons, see the discussion of "group bias" in chapter 8.

18. *EE*, 241.

19. See for example the works of Dorceta E. Taylor, *Toxic Communities: Environmental Racism, Industrial Pollution, and Residential Mobility* (New York: New York University Press, 2014) and *The Rise of the American Conservation Movement: Power, Privilege, and Environmental Protection* (Durham, NC: Duke University Press, 2016). See also Bruce E. Johansen, *Environmental Racism in the United States and Canada: Seeking Justice and Sustainability* (Santa Barbara, CA: Praeger, 2020), and Luke W. Cole and Sheila R. Foster, *From the Ground Up: Environmental Racism and the Rise of the Environmental Justice Movement*, Critical America 34 (New York: New York University Press, 2001).

20. *EE*, 256.

21. *EE*, 228.

22. *EE*, 260.

23. *EE*, 260.

24. See chapter 8.

25. *Insight*, 7–8.

26. See the appendix, "A Method for Environmental Ethics."

Chapter 7: Toward the Wholeness of the Emerging Good

1. For further details, see the appendix, "A Method for Environmental Ethics." Robert Doran identified a fourth "psychic" conversion. See *EOD*, 237–38.

2. *Insight*, 656.

3. *Insight*, 472.

4. *Insight*, 197. Lonergan uses this passage to characterize the self-correcting process in commonsense knowing, but it also characterizes his understanding of scientific knowing. See also chapter 1 of this book.

5. *Insight*, 320.

6. "My intention was an exploration of methods generally in preparation for a study of the method of theology." Bernard Lonergan, "Insight Revisited," in *A Second Collection*, Collected Works of Bernard Lonergan, vol. 13, ed. Robert M. Doran and John D. Dadosky (Toronto: University of Toronto Press, 2016), 226. Cited hereafter as *Second Collection*.

7. Lonergan actually developed his account of emergent probability twice. The first time it was built upon his analysis of the four scientific methodological patterns. The second time it was built upon the deeper cognitional structure that is utilized in all of human thinking. Compare *Insight* 138–51 with 460–76.

8. *Insight*, 27; see also Aristotle, *Posterior Analytics* II.1 89b27–29.

9. *Insight*, 426 and 386, respectively, emphasis added. Lonergan said this of the "integral heuristic structure" of his method of metaphysics, which is a more sophisticated version of empirical probability.

10. For present purposes I must limit my discussion to only the first two—classical and statistical methods—of the four patterns that Lonergan identified, and how these two contributed to his notion of emergent probability. "Genetic method" is the third of these general methodical patterns and is used to investigate development as such in sciences such as embryology and developmental psychology. "Dialectic" is the fourth method and is discussed elsewhere in this book. Lonergan use the term *generalized emergent probability* for the heuristic of emergent wholeness that incorporates all four methods, but he reserved emergent probability for the worldview that takes into account only classical and statistical methods. For the sake of simplicity, I am discussing only emergent probability, although most of the ethical implications also hold for generalized emergent probability. For Lonergan's argument that there are four such patterns, see *Insight*, 509–10.

11. *Insight*, 61–63. Lonergan's meaning of "correlation" is quite different from what is commonly called a "merely statistical correlation."

12. *Insight*, 88, 143–44, 148.

13. *Insight*, 141.

14. *Insight*, 112–13, 131.

15. *Insight*, 131.

16. *Insight*, 144.

17. *EE*, 210.

18. *EE*, 212.

19. Lonergan used the technical term *counterposition* to denote untenable assumptions that are performative contradictions. See chapter 8.

20. *Insight*, 413–14.

21. *Insight*, 416. For complete details, see 456–76.

22. See chapter 1. See also Bernard Lonergan, *Verbum: Word and Idea in Aquinas*, ed. Frederick E. Crowe and Robert M. Doran, Collected Works of Bernard Lonergan, vol. 2 (Toronto: University of Toronto Press, 1997), 51; and Patrick H. Byrne, "Desiring and Practical Reason: MacIntyre and Lonergan," *International Philosophical Quarterly* 60, no. 1, issue 237 (March 2020): 75–96.

23. *Insight*, 629.

24. See *Insight*, 628–30. For difficulties with Lonergan's argument, and an expanded argument, see *EOD*, 358–70.

25. *Insight*, 629.

26. See *EOD*, 207–40.

27. It is a rare person who even tacitly takes into account all the conditions that a fully objective deliberation would call forth. Environmental ethics has been awakening people to conditions and consequences they have ignored for a long time.

28. For the fuller details of the argument and how it can be made stronger, see *EOD*, 358–70.

29. *Insight*, 238.

30. *Insight*, 629.

31. *Insight*, 238.

32. See *EOD*, 227–34.

33. MT, 32.

34. See *EOD*, 241–84, 358–86.

35. *Insight*, 254–57.

36. *EE*, 161.

37. *Insight*, 470.

38. See *EOD*, 227–37.

39. *Insight*, 244–67.

40. *Insight*, 255.

41. Lonergan argued that the forces of prejudice, bias, and ignorance overwhelm human beings so deeply that something even more than moral and intellectual conversion is needed. He called this "religious conversion." He did not mean conversion to some religious organization but a conversion to the consciousness of unconditional love that is dimly present in the consciousness of every human being. See *MT*, 104–18, 237–44, and *EOD*, 226–27.

Chapter 8: The Dialectic of Environmental Ethics

1. *EE*, 255, emphasis added. DesJardins does not discuss "consequentialism" or its role in the history of environmental ethics. Consequentialism is a

somewhat later revision of utilitarianism. Instead of pleasures, pains, or preferences, consequentialism seeks to maximize other kinds of goods or values. Within consequentialism there is much debate about what should count as the goods or values to be substituted into the calculus. In chapter 7 and *EOD* chapters 9 and 11–14 I have presented an alternative way of conceiving of the wholeness of the good, including values and a ground for objectively evaluating differing values.

2. For further details, see the appendix, "A Method for Environmental Ethics."

3. Bernard Lonergan, "Cognitional Structure," in *Collection*, Collected Works of Bernard Lonergan, vol. 4, ed. Frederick E. Crowe and Robert M. Doran (Toronto: University of Toronto Press, 1988), 217. Hereafter cited as CS.

4. *Insight*, 443–44.

5. For further details, see chapter 7 and the appendix, "A Method for Environmental Ethics."

6. *Insight*, 413.

7. *EE*, 13, 31, 170–71.

8. See "Desiring and Practical Reason: MacIntyre and Lonergan," *International Philosophical Quarterly* 60, no. 1, issue 237 (March 2020): 75–96. See also Bernard Lonergan, "The Form of Inference," in *Collection*, Collected Works of Bernard Lonergan, vol. 4, ed. Frederick E. Crowe and Robert M. Doran (Toronto: University of Toronto Press, 1988), 3–16.

9. See *EOD*, 95–117.

10. *EE*, 78.

11. *EE*, 78–79.

12. Lonergan elaborated the scale of values pertaining to human beings via his "structure of the human good." In that structure, the lowest level has to do with the vital values that satisfy "particular goods"—which is to say, a momentary instance of a vital value that satisfies a particular, ephemeral need. For his full discussion of the structure of the human good, see *MT*, 47–51. See also *EOD*, 309–30.

13. *Insight*, 629.

14. "But empiricism amounts to the assumption that what is obvious in knowing is what knowing obviously is." *Insight*, 441.

15. *EE*, 34.

16. As was mentioned in chapter 2, he insisted that the resources of the forest should benefit all citizens, not just the wealthy few. *EE*, 53.

17. *EE*, 140.

18. *EE*, 145.

19. *MT*, 273, 248, respectively.

20. *EE*, 256.

21. See chapter 6.

22. *EE*, 256.
23. See *EE*, 259.
24. *Insight*, 8.
25. *EE*, 256.
26. *EE*, 261.
27. *Insight*, 414, emphasis added.
28. *EE*, 255.
29. *EE*, 161.
30. *EE*, 248.
31. See *EE*, 181.
32. *Insight*, 472.
33. *Insight*, 498.
34. *Insight*, 245. Lonergan's words were used in the context of his discussion of individual bias but apply equally to group bias.
35. See *EE*, 98, 151–52. See also *Laudato si'*, §§115–22, 137–62.
36. *Insight*, 498.

Chapter 9: The Rise of Uniformitarianism

1. DesJardins does begin his book with a brief discussion exercise about the climate deniers' challenges to climate science (*EE* 3–6), but this primarily serves to introduce the epistemological is/ought problem, and there is no historical narrative comparable to the other periods covered in his book.

2. James Rodger Fleming, *Historical Perspectives on Climate Change* (New York: Oxford University Press, 1998); Tobias Krüger, *Discovering the Ice Ages: International Reception and Consequences for a Historical Understanding of Climate*, trans. Ann M. Hentschel (Boston: Brill, 2013); Spencer R. Weart, *The Discovery of Global Warming*, rev. ed. (Cambridge, MA: Harvard University Press, 2008); Nathaniel Rich, *Losing Earth: A Recent History* (New York: Farrar, Straus and Giroux, 2019). Weart's book also includes a supplementary website containing two dozen essays and seven hundred additional hyperlinks and references to over a thousand scientific and historical publications. In addition, for certain facts regarding the rise of uniformitarianism, I have relied upon Loren Eisley's *Darwin's Century: Evolution and the Men Who Discovered It* (Garden City, NY: Doubleday & Company, 1958). There are numerous other histories of environmental and climate change thought that are not explicitly referenced in this book.

3. Weart, ix.

4. Fleming, viii. Spencer Weart also underscores the importance of the history of climate change science in order to "understand our situation by explaining how we got here" (viii.)

5. Fleming, 33, 130.

6. Thomas S. Kuhn, *The Structure of Scientific Revolutions*, second ed., enlarged (Chicago: University of Chicago Press, 2012), 43ff.

7. Fleming, 5. Although in his introduction Fleming writes that the notion of "apprehension" was the metaphor that helped him organize his narrative, the idea of "privileged position" keeps recurring throughout the book.

8. Rich, 188–89. Although the 1950s is mentioned repeatedly in the passage where Rich catalogues the "everyone" who knew by this point, in several cases the facts he cites occurred decades later.

9. See chapter 15.

10. Weart, 156–57.

11. Weart, 165.

12. Fleming, 5.

13. Fleming, 11–19.

14. Fleming, 23; see also Richard Slotkin, *Regeneration through Violence: The Mythology of the American Frontier, 1600–1860* (Norman: University of Oklahoma Press, 1973), 25–93.

15. Fleming, 24.

16. Fleming, 24, 28–29.

17. Fleming, 25.

18. Fleming, 31.

19. Fleming, 37, also 41.

20. Fleming, 44.

21. Fleming, 33–44; see also his discussion of T. C. Chamberlain at a somewhat later time, 85–86.

22. Fleming, 49. Forry did admit that localized efforts of land clearance and urban growth had some effects on climate, but that these were limited in area and minimal compared with far more influential factors such as latitude.

23. For example, Lorin Blodget, Elias Loomis, H. A. Newton, Charles A. Schott, Cleveland Abbe, and especially William Ferrel. Fleming, 50–53.

24. Fleming, 53.

25. In 1815 he finally published his geological map covering England, Wales, and parts of Scotland. But his findings had become widely recognized many years earlier. Eisley, 75–81.

26. Eisley, 65–69.

27. Eisley, 69–74.

28. Eisley, 85.

29. Dov Ospovat, *The Development of Darwin's Theory: Natural History, Natural Theology, and Natural Selection, 1838–1859* (New York: Cambridge University Press, 1981), 56.

30. Charles Lyell, *Principles of Geology. Or, The Modern Changes of the Earth and Its Inhabitants Considered as Illustrative of Geology*, ninth ed. (New York: D. Appleton, 1854), ix (beginning of the table of contents).

31. Eisley, 99.
32. Eisley, 103–04.

Chapter 10: The Anomaly of Ice Ages

1. Krüger, 33, 61.
2. Krüger, 77.
3. Krüger, 442.
4. Krüger, 26–33, 62.
5. Krüger, 34–36.
6. Krüger, 37–43, 70–78, 89.
7. Imre Lakatos, "The Methodology of Scientific Research Programmes," in *Philosophical Papers*, vol. 1, ed. J. Worrall and G. Curries (Cambridge: Cambridge University Press, 1978).
8. Krüger, 149.
9. Krüger, 62.
10. Krüger, 97.
11. Krüger, 103.
12. Krüger, 143.
13. Krüger, 103–04.
14. Krüger, 148–49; see also 103.
15. Krüger, 152.
16. Krüger, 180; see also chapter 11.
17. Some decades later Lord Kelvin performed advanced calculations, arguing that the earth would have been far too hot to support life just three hundred thousand years ago. He explicitly intended this calculation to raise a serious scientific problem for Darwin's theory of evolution.
18. Krüger, 153.
19. Krüger, 157.
20. Krüger, 160.
21. Krüger, 165.
22. Krüger, 177.
23. Krüger, 170. Krüger offers a lengthy account of the likely contributions of each to the ultimate synthesis, 167–70.
24. Krüger, 187–88.
25. Krüger, 177–78.
26. Krüger, 178.
27. Krüger, 186.
28. Which Krüger documents, 182.
29. Krüger, 184–85.
30. Krüger's chapter would be a rich source for Dialectical analysis of the sort offered in chapter 8 to see where biases and conversions played roles in

the gradual acceptance of ice ages. Such an analysis, however, would be too great a distraction to the overall objectives of this book and must be left for some other time.

31. Krüger, 455–56.

32. Krüger, 460.

33. For Lonergan's meaning of judgments as "virtually unconditioned," see *Insight*, 305–24.

Chapter 11: Joseph Fourier and the Science of Heat Dynamics

1. Fleming, 9.

2. Joseph Fourier, *The Analytic Theory of Heat*, trans. Alexander Freeman (London: At the [Cambridge] University Press, 1878), 1–2.

3. As quoted in Fleming, 63.

4. Fleming, 65.

5. *Insight*, 318–19 and 357–58.

6. As quoted by Fleming, 55; see also 61.

7. Here I have used the literary device of three approximations to communicate the basic ideas. Fourier himself did not use these terms or proceed in exactly these steps.

8. As quoted by Fleming, 61.

9. As quoted by Fleming, 61.

Chapter 12: The Growth of the Science of Atmospheric Warming

1. Weart, 15.

2. Weart, 13–15. Abbot's work came decades later, beginning around 1913.

3. Krüger, 408–16; Weart, 16, 48–49.

4. Weart, 48.

5. Weart, 16.

6. Fleming, 67.

7. Fleming, 73, 70.

8. Fleming, 73.

9. Fleming, 79.

10. Fleming, 76.

11. *Insight*, 61–62.

12. Fleming, 76.

13. Weart, 5.

14. Fleming, 76.
15. Fleming, 80.
16. Fleming, 81.
17. Fleming, 81.
18. Fleming, 82.
19. Fleming, 78.
20. Fleming, 77, 79.
21. Fleming, 86.
22. For details, see Fleming, 87–90.
23. Fleming, 93.
24. Arrhenius and Chamberlin had relied upon "interpolations" from the measurements by Samuel Langley in 1881. Fleming, 73, 77.
25. Fleming, 107.
26. Fleming, 109.
27. Fleming, 90.
28. Fleming, 107.
29. Fleming, 114.
30. Fleming, 116.
31. Fleming, 117–18.
32. Fleming, 115.
33. Fleming, 119–121.
34. Gilbert N. Plass, "The Carbon Dioxide Theory of Climate Change," *Tellus* 8, no. 2 (1956): 140–54, accessed at https://onlinelibrary.wiley.com/doi/pdf/10.1111/j.2153-3490.1956.tb01206.x 7/31/2020.
35. Fleming, 149.
36. Fleming, 121.
37. Fleming, 122.
38. See online blog "The Carbon Dioxide Theory of Gilbert Plass," *Real-Climate: Climate Science from Climate Scientists*, of Gavin A. Schmidt, climate modeler at the NASA Goddard Institute for Space Studies and Earth Institute at Columbia University in New York. Gavin Schmidt, January 2010, accessed at http://www.realclimate.org/index.php/archives/2010/01/the-carbon-dioxide-theory-of-gilbert-plass/ 7/31/2020.
39. As quoted in Fleming, 122, emphasis added.
40. Weart, 25.
41. Fleming, 125.
42. Weart, 27.
43. As quoted by Fleming, 125, from Roger Revelle and Hans E. Suess, "Carbon Dioxide Exchange between Atmosphere and Ocean and the Question of an Increase in Atmospheric CO_2 during the Past Decades," *Tellus* 9 (1957): 18–27.
44. Fleming, 125.

45. Weart, 19–37.

46. The annual downswings are due to seasonal cycles that absorb CO_2.

47. Fleming, 126.

48. Weart, 42, emphasis added; 4°C is actually equivalent to 7.2°F.

49. Weart, 43, emphasis added.

Chapter 13: The Rise of Computer Modeling

1. Fleming, 130.

2. See especially Weart, 53–57, 92–97, 111, 114–22, 127–36.

3. Weart, 54.

4. Weart, 54.

5. Weart, 56.

6. Weart, 58–59.

7. Weart, 59.

8. As quoted in Weart, 59.

9. James Gleick, *Chaos: The Making of a New Science* (New York: Viking Penguin, 1987), 11–31.

10. Lorenz was the originator of the famous quote: "Does the flap of a butterfly's wing in Brazil set off a tornado in Texas?" Weart, 114.

11. Weart, 115.

12. *Insight*, 133.

13. Weart, 45–47.

14. Weart, 51.

15. Weart, 75, 127.

16. Weart, 129.

17. Spray can use of CFCs was banned in the US and some other countries in 1978 because they eroded the ozone layer, allowing dangerous wavelengths of ultraviolet to penetrate to the earth's surface. By 1995 strict regulations to recover CFCs from air conditioning units during servicing or disposal were put in place.

18. Weart, 116–18.

19. Weart, 116–18.

20. Weart, 96–97.

21. Weart, 132–33.

22. Weart, 134.

23. Weart, 105–06.

24. Weart, 107. It is noteworthy that prior to considering the relationships between plant and animal ecosystems and climate change, the horizon for ethical thought was life-as-a-whole but not yet the integral whole of all that is good—what I called "the emerging good" or Pope Francis has called "integral ecology."

25. Weart, 106–12.

26. Weart, 99, 102, emphasis added.
27. Weart, 99.
28. Weart, 65.
29. Weart, 96.
30. Weart, 61.
31. Weart, 61.
32. Weart, 61–62.
33. Weart, 75.
34. Weart, 99, 101.
35. Weart, 146.
36. See chapter 15.
37. Weart, 118.
38. Weart, 137.
39. Weart, 149. The full title of the conference was the World Conference on the Changing Atmosphere: Implications for Global Security, nicknamed the "Toronto Conference."
40. Weart, 156, 164, 178.
41. Weart, 193.

Chapter 14: Going Public: Ethical Response to Climate Science

1. *EOD*, 99–103.
2. Regarding probable truth and ethical responses to it, see chapter 16.
3. Rich, 4, and https://www.nrdc.org/stories/ipcc-climate-change-reports-why-they-matter-everyone-planet?gclid=Cj0KCQjwmN2iBhCrARIsAG_G2i586CzIOTy6rrPyN2okPyx2C3rpXLS3AqTldS7_weqe_hYDB_kv-2oaAg5ZEALw_wcB#sec-whatis, accessed May 7, 2023.
4. In its 2023 report, the IPCC predicts that global average temperature will rise to 1.4°C–2.7°C above the level at the onset of the Industrial Revolution. See https://report.ipcc.ch/ar6syr/pdf/IPCC_AR6_SYR_LongerReport.pdf. This IPCC report came after the historical period under discussion in this section.
5. Source for the figures in items (10) is the UCAR Center for Science Education website, https://scied.ucar.edu/longcontent/predictions-future-global-climate, accessed August 15, 2020. Weart uses the figure of "3°C rise—more or less" (195).
6. Weart, 194. Other sources put the estimated extinction rate higher.
7. Weart, 194.
8. *MT*, 32–33. See also *EOD*, 246–84, for a fuller discussion.
9. On the emerging good, see chapter 7.
10. Weart, 86.

11. Integration of interdisciplinary collaborations is one of the objectives of Lonergan's heuristic method of emergent probability and its extension into the heuristic of the emerging good. See chapter 7.

12. *Insight*, 236–37.

13. *Insight*, 197–200.

14. Fleming, 119.

15. Today, however, all these claims are now well substantiated.

16. Weart, 42.

17. Fleming, 123.

18. Fleming, 128.

19. Weart, 41.

20. The chemical industries tried mightily to discredit her and her book.

21. Weart, 39–41.

22. Weart, 69.

23. Weart, 86.

24. Weart, 69.

25. Weart, 151.

26. Weart, 87.

27. *Insight*, 251.

28. See chapter 13.

29. Weart, 112.

30. Weart, 113.

31. Weart, 150.

32. Weart, 150. Weart does not make clear who used the term *super hurricane* or to which storm it refers. The most likely candidate would be Hurricane Gilbert, which impacted Jamaica on September 12, 1988, and then struck the Yucatan Peninsula at Category 5 strength on September 13. Gilbert killed 318 people and caused about $2.98 billion in damages. However, since this did not strike the US mainland and would have made less impact on the consciousness of the average US citizen, it is also possible that the phrase refers to Hurricane Florence, which struck the Mississippi Delta on September 10, 1988 (not to be confused with Hurricane Florence of 2018). Although there was only one reported fatality and damage estimated in the vicinity of $3 million across Louisiana, Alabama, and the Florida Panhandle, prior to landfall there had long been great worry about a storm that would hit New Orleans whose storm surge would overwhelm the flood protection systems in that city. This of course did happen in 2005 with Hurricane Katrina. Perhaps more than any other storm, Hurricane Sandy of 2012 has become the icon of climate change superstorms. Its devasting impacts on New York City received dazzling media attention.

33. Weart, 149–50, emphasis added.

34. Weart, 152–53.

35. Weart, 156.

36. "Computer modelers . . . remained at the convergence of many research programs." Weart, 190.

37. Weart, 165.

38. Weart, 178.

39. *Insight*, 188.

40. While the range of interest on the part of modern empirical science is wider than that of common sense, it too imposes limitations on its field of investigations. It is limited to questions and insights that require sense data as part of the conditions for verification. This excludes other questions and insights that pertain to possibilities that lie outside of the sensible realm, for example, questions and insights about religious and transcendent possibilities, as well as questions and insights that pertain to what Lonergan called the data of consciousness (versus the data of sensation). *Insight*, 260–61 and 657–58.

41. Weart, 151.

42. *MT*, 78–81.

43. *MT*, 81–82, 110–11, 241–43.

44. *MT*, 80–82.

45. *MT*, 82.

46. John Sununu tried to do just this while he was chief of staff to President George H. W. Bush. On the basis of his master's degree in engineering, he arrogantly thought himself competent to do a computer model of climate on his desktop computer. See chapter 15.

47. *Insight*, 251.

48. *Insight*, 69–70.

49. Weart, 151.

Chapter 15: The Dialectic of Politics and Climate Change Science

1. Sources: https://www.epa.gov/energy/electricity-customers and https://www.epa.gov/ghgemissions/sources-greenhouse-gas-emissions, respectively, accessed August 16, 2020.

2. Weart, 188.

3. Nathaniel Rich, *Losing Earth: A Recent History* (New York: Farrar, Straus and Giroux, 2019, 47. Cited hereafter as Rich.

4. Rich, 81–83.

5. Jonathan Sacks, *To Heal a Fractured World* (New York: Schocken Books, 2005), 37–38. Sacks explains that the tradition is given a classic summary by Maimonides.

6. Rich, 51.

7. In his bibliography, Rich lists the studies by both Fleming and Weart as among his sources.

8. Rich, 6.

9. Rich, 7.

10. Rich, 13.

11. Rich, 17–18.

12. The report was entitled "Carbon Dioxide and Climate: A Scientific Assessment."

13. Rich, 36–37.

14. Rich, 66–67.

15. Rich, 70.

16. Rich, 76–77.

17. Rich, 90, emphasis added.

18. Rich, 90–91.

19. Rich, 91.

20. *NYT*, October 21, 1983, 1. See also Rich, 94.

21. Weart, 150.

22. Rich, 102.

23. Rich, 109–10.

24. Rich, 113.

25. Rich, 110.

26. Rich, 106.

27. Rich, 112.

28. *Insight*, 250–51, 319, and 412–13. Regarding the "counterposition" and the differentiation of scientific from commonsense consciousness, see chapters 8 and 14.

29. Rich, 117. Rich later reveals that the "nameless censor in the White House hiding behind the OMB letterhead" was Fred Koomanoff, program director of the Carbon Dioxide Research Division at the Department of Energy during the Reagan administration (153).

30. Rich, 119.

31. As quoted in Rich, 131–32.

32. Rich, 132.

33. Rich, 137–38.

34. Rich, 149.

35. Rich, 149–50.

36. Rich, 147–48.

37. Rich, 151.

38. Rich, 152.

39. Rich, 162.

40. Rich, 153.

41. Rich, 158–59.

42. Rich, 162.

43. Riley E. Dunlap and Aaron M. McCright, "Organized Climate Change Denial," in *The Oxford Handbook of Climate Change and Society*, ed. John S. Dryzek, Richard B. Norgaard, and David Schlosberg (New York: Oxford University Press, 2018), from the Oxford Handbooks Online (www.oxfordhandbooks.com). Cited hereafter as Dunlap and McCright.

44. Weart, 158.

45. Rich, 182.

46. Dunlap and McCright, 1.

47. For details, see Dunlap and McCright, 5–14.

48. See the appendix, "A Method for Environmental Ethics."

49. Arlie Russell Hochschild, *Strangers in Their Own Land: Anger and Mourning on the American Right* (New York: New Press, 2016), 10ff. Cited hereafter as *Strangers*.

50. *Strangers*, 79–80.

51. *Strangers*, 79.

52. *Strangers*, 68.

53. *Strangers*, 135.

54. See the appendix, "A Method for Environmental Ethics."

55. *Strangers*, 136–40, emphasis added. Compare Hochschild's symbolism of Line-Cutting with Isabel Wilkerson's image of securing one's place on the caste ladder. See chapter 11, Isabel Wilkerson, *Caste: The Origins of Our Discontents* (New York: Random House, 2020). Cited hereafter as *Caste*.

56. *Strangers*, 144 and 140, respectively.

57. *Strangers*, 190.

58. For references to these works, see the list at the end of this chapter.

59. *Caste*, 40.

60. *Strangers*, 183.

61. *Strangers*, 236.

62. See the appendix, "A Method for Environmental Ethics."

63. *EOD*, 238–40.

64. See *EOD*, chapters 6, 7, and 8.

Chapter 16: Probability, Uncertainty, and Predictability in Science

1. *Caste*, 17, 236.

2. Dunlap and McCright, 1.

3. As quoted by Rich, 111.

4. *Second Collection*, 88.

5. *Insight*, 324.

6. *MT*, 223.

7. Lonergan's way of drawing the distinction is the subject of this chapter. Regarding Popper's approach, see Patrick H. Byrne, "Statistics as Science: Lonergan, McShane, and Popper," *Journal of Macrodynamic Analysis* 3 (2003): 55–75, http://www.mun.ca/jmdavol3/byrne.pdf. Regarding Boyle, see Steven Shapin and Simon Schaffer, *Leviathan and the Air Pump: Hobbes, Boyle and the Experimental Life* (Princeton, NJ: Princeton University Press, 2011), 2–24. John Locke was also influential in the standard view of the probability of propositions. He cites the conformity of the proposition to our knowledge, observation and experience, and the testimony of others: "The grounds of probability are two: conformity with our own experience, or the testimony of others' experience. Probability then, being to supply the defect of our knowledge and to guide us where that fails, is always conversant about propositions whereof we have no certainty, but only some inducements to receive them for true." *An Essay Concerning Human Understanding* IV.XV.4.

8. Although averages and ideal frequencies are closely related, there are also important differences. Averages are more familiar and are commonly used as stand-ins for ideal frequencies, so I will often use "average" as a shorthand to keep explanations simple.

9. *Insight*, 81–82.

10. "If now the results of the whole of the experiments be brought together, there is found . . . an average ratio of 2.98 to 1, *or 3 to 1*." Gregor Mendel, "Experiments in Plant Hybridization," 11, emphasis added, http://www.esp.org/foundations/genetics/classical/gm-65.pdf, accessed October 24, 2021.

11. *Insight*, 86.

12. In chapter 9 we saw how data gathering alone was found to be an inadequate "privileged position" for climate science.

13. CS, 218.

14. *Insight*, 305–10.

15. CS, 218.

16. *Insight*, 309.

17. *Insight*, 309–10.

18. *Insight*, 197.

19. *Insight*, 197.

20. *Insight*, 325–26. Of course, postmodern thinkers from Kuhn to Foucault to Lyotard dispute this claim that the self-correcting process approaches any limit. Because a full response to this set of objections is not possible in this work, here I can do no more than point out that the phenomenon of questioning itself is preconceptual, prior to, and the basis of all paradigms, epistemes, and metanarratives. Questioning sets a standard to which paradigms, epistemes, and metanarratives attempt to respond, and which goes beyond them by calling

their imperfections into question. For fuller discussion, see Patrick H. Byrne, "Culturally Situated Scientific Reason and Objectivity," *Proceedings from the Institute for Liberal Studies Conference on Science and Culture* 10 (1999): 17–22.

21. Weart, 118.

22. Weart, viii.

23. Weart, 65.

24. Fleming, 119–21.

25. Weart, 85.

26. Weart, 47.

27. Simone Pierre de Laplace, A *Philosophical Essay Concerning Probabilities*, trans. Frederick Wilson Truscott and Frederick Lincoln Emory (New York: Dover, 1952; originally published in French 1814), 4, emphasis added. Cited hereafter as Laplace.

28. Laplace, 6.

29. *Insight*, 90.

30. *Insight*, 76.

31. Regarding one aspect of the theoretical impossibility, see Patrick H. Byrne, "Relativity and Indeterminism," *Foundations of Physics* 11 (1981): 913–32.

32. Weart, 202.

33. Weart, 201–02, emphasis added.

34. Weart, 157.

Chapter 17: Scientific Consensus, Trust, and Belief

1. Weart, 197, emphasis added.

2. See *Insight*, 725–41.

3. Weart, 156.

4. *MT*, 43.

5. Bernard Lonergan, "Belief: Today's Issue," in *Second Collection*, 76.

6. *MT*, 43.

7. Weart, 198–99.

8. *MT*, 42.

9. Weart, 199, emphasis added.

10. See, for example, Barry Barnes, David Bloor, and John Henry, *Scientific Knowledge: A Sociological Analysis* (Chicago: University of Chicago Press, 1996), and Vivien Burr, *Social Constructionism*, third ed. (New York: Routledge, 2015).

11. Weart, 200, emphasis in the original.

12. *MT*, 45–46.

13. *MT*, 46.

14. Weart, 177–78, emphasis added.

15. Weart, 199, emphasis added.

16. *MT*, 43.
17. Weart, 197–98.
18. Weart, 199.
19. Weart, 198–201, emphasis added.
20. Katharine Hayhoe, *Saving Us: A Climate Scientist's Case for* Hope *and* Healing *in a Divided World* (New York: One Signal, 2021), xi–xii. Cited hereafter as Hayhoe.
21. Hayhoe, 31–32.
22. Hayhoe, 18.

Chapter 18: *Laudato si'* and Integral Ecology

1. See the appendix, "A Method for Environmental Ethics."
2. As quoted by Nathan Schneider, "The Church and the World: A Prehistory of *Laudato si'*," *America*, July 14, 2015. McDonagh's comment points to the need for further work on the important contributions of all religions, not only Roman Catholicism, to the history of environmental ethics and action. Nevertheless, *Laudato si'* has had a profound influence that extends well beyond the Roman Catholic Church.
3. See chapter 7. See also *Insight*, 416–19. In *Insight*, the integral heuristic structure is intended to guide collaboration among scientists and people of common sense in progressing toward knowledge of the natural world (understood as including human beings, their actions and constructions). It is especially intended to facilitate cooperative and interdisciplinary work that explores the dynamic, complex interrelationships and dependencies between the human and nonhuman dimensions of nature. Lonergan's subsequent account of the "ontology of the good" (*Insight*, 628–30; see also *EOD*, 309–86) provides a parallel heuristic for collaboration in knowing and actualizing what is good in all its dynamic and mutually interdependent complexities. I have expanded Lonergan's notion of the ontology of the good in my account of the heuristic of the emerging good.
4. *Insight*, 60–61, 337–38.
5. Neil Ormerod and Cristina Vanin, "Ecological Conversion: What Does It Mean?," *Theological Studies* 77, no. 2 (2016): 328–52. Hereafter cited as Ormerod and Vanin.
6. Ormerod and Vanin, 330.
7. Lynn White Jr., "The Historical Roots of Our Ecological Crisis," *Science*, New Series, 155, no. 3767 (March 10, 1967): 1205. Cited hereafter as White.
8. *MT*, 112.
9. See the appendix, "A Method for Environmental Ethics," §7.2.
10. *MT*, 104.
11. *MT*, 103.

12. *MT*, 112. See also Ormerod and Vanin, 333.

13. Ormerod and Vanin, 336–44. Unfortunately, they do not expand the connection among the human scale of values—vital, social, cultural, personal, and religious—into what Lonergan called the "ontology of the good" and I have called the emerging good that encompasses the good of physical, chemical, organic, and animal values as well. See the appendix, "A Method for Environmental Ethics," §7.2.

14. Ormerod and Vanin, 344.

15. *MT*, 315.

16. White, 1207.

17. On the wholeness of being as beloved and calling forth loving response, see Anthony J. Steinbock, *Knowing by Heart: Loving as Participation and Critique* (Evanston, IL: Northwestern University Press, 2021), 33–42.

18. See the appendix, "A Method for Environmental Ethics."

19. *MT*, 96–98.

20. *Insight*, 523–26, 657–99.

21. *Insight*, 679.

22. "In this matter of distinguishing a theology of the people from Marxism, it is important to recognize the contribution made to the theology of the people by Rafael Tello." Gerard Whelan, SJ, *A Discerning Church: Pope Francis, Lonergan, and a Theological Method for the Future*, Theology at the Frontiers (New York: Paulist Press, 2019), 100. Cited hereafter as Whelan.

23. Whelan, 139.

24. Whelan, 101.

25. Whelan, 3.

26. Whelan, 1–4, 111–12.

Chapter 19: The United Nations Strategic Goals for Sustainable Development

1. Rich, 193.

2. Austen Ivereigh, "Why Call It Progress?," *Commonweal*, September 30, 2019.

3. Jeffrey D. Sachs, *The Age of Sustainable Development* (New York: Columbia University Press, 2015). Cited hereafter as Sachs. See also "History" at https://sdgs.un.org/goals.

4. Sachs, 284.

5. Sachs, 7–8.

6. Sachs, 12–13.

7. Sachs, 49–66.

8. Carson, 6–7.

9. *Insight*, 478–79.

10. Sachs, 67.

11. *The Sustainable Development Goals Report 2016*, United Nations, https://unstats.un.org/sdgs/report/2016/The%20Sustainable%20Development%20Goals%20Report%202016.pdf. See also *Global Indicator Framework for the Sustainable Development Goals and Targets of the 2030 Agenda for Sustainable Development*, https://unstats.un.org/sdgs/indicators/Global%20Indicator%20Framework%20after%20refinement_Eng.pdf, and https://socialimpactmovement.org/the-complete-list-of-sustainable-development-goals/?gclid=CjwKCAjwmqKJBhAWEiwAMvGt6MJnVa9VN5QIqCdSLoSJL8Ubl9u2hkgwr0AdfKZ8Vjs-NBJZm6xbMBoCQdwQAvD_BwE, all three accessed September 27, 2021.

12. Found at https://sdg-tracker.org/energy, accessed September 27, 2021.

13. See https://sdgs.un.org/goals/goal1, accessed September 27, 2021.

14. Sachs, 102–03.

15. The seven factors are poverty traps, inappropriate economic policies, government insolvency, landlocked geography, poor governance, cultural barriers, and military insecurity. See Sachs, 103–09, for a detailed discussion.

16. Sachs, 175–80 and 484, respectively.

17. *Insight*, 259.

18. However, the language of the earlier draft to "phase out" coal power and inefficient fossil fuel subsidies was blocked by India (and a few others would have if India had not relieved them of the task) because of India's need to catch up in economic development. The language of any official document of a COP must be agreed to by all 196 nations.

19. For the full text of the agreements, see https://unfccc.int/process-and-meetings/conferences/glasgow-climate-change-conference-october-november-2021/outcomes-of-the-glasgow-climate-change-conference, accessed August 20, 2023.

Conclusion

1. Weart, viii.

Appendix: A Method for Environmental Ethics

1. *EOD*, 413–48.

2. *MT*, 337. Lonergan's actual terms are policymaking, planning, and execution.

3. *MT*, 129–30.

4. CS, 207.

5. CS, 215.

6. *Insight*, 346–48.

7. CS, 211–14.

8. For full details, see *EOD*, 95–117.

9. *MT*, 31–32. See also *EOD*, 136–81.

10. Lonergan contributed some of his own innovations to this specialty. See *MT*, 145–65.

11. *MT*, 164–219.

12. For details, see *Insight*, 214–20, 241–51.

13. *Insight*, 244, 251.

14. *MT*, 272–73, 269, respectively.

15. *MT*, 238.

16. See *MT*, 237.

17. *MT*, 364.

18. *MT*, 365.

19. *MT*, 336.

20. Ethical principles have received such formulations as: "Act virtuously"; "Act to maximize pleasure (and minimize pain)"; "Act to treat humanity as an end and never simply as a means"; "We must preserve the environment"; "A true ecological approach must integrate questions of justice in debates on the environment."

21. *MT*, 131–32 and 142; see also *EOD*, 442–43.

22. *MT*, 267.

23. *MT*, 282–88.

24. Plato, *Republic* VII, 518c.

25. "Here by reason I would understand the compound of the activities on the first three levels of cognitional activity, namely, of experiencing, of understanding, and of judging." *MT*, 111–12. See also Patrick H. Byrne, "Desiring and Practical Reason: MacIntyre and Lonergan," *International Philosophical Quarterly* 60, no. 1, issue 237 (March 2020): 75–96.

Index